Skin Care for Cold Weather

Skin Care for Cold Weather

By

Alexandra Bonsall

Cambridge Scholars Publishing

Skin Care for Cold Weather

By Alexandra Bonsall

This book first published 2019

Cambridge Scholars Publishing

Lady Stephenson Library, Newcastle upon Tyne, NE6 2PA, UK

British Library Cataloguing in Publication Data
A catalogue record for this book is available from the British Library

ISBN (10): 1-5275-3946-6
ISBN (13): 978-1-5275-3946-4

For Eva and Rory.

The author has made every effort to ensure that the information in this book is correct and up to date as of June 2019, but medical advice, practice and laws may change. The information in this book has been compiled from the author's knowledge and reliable source texts with sources referenced throughout. It is intended as general guidance on subjects included, not as a substitute for medical advice and should not be relied upon for medical, pharmaceutical or professional advice on an individual basis. The reader should consult a qualified health professional for up to date advice on matters relating to his/her health.

The author and publishers do not assume, and disclaim as far as the law allows, any liability to any party for any loss, damage or disruption caused directly or indirectly from the use or misuse of the information contained in this book and from any errors or omissions, whether occurring by accident or other cause. The author has not received any financial incentives for recommendations made in this book.

Photography

The very talented Catherine Davidson of Catherine Davidson Photography, Westhill, Aberdeen, kindly took the image for the front cover. Other images within the text are the author's own.

Contents

Introduction

Aberdeen, Scotland 57.15° North, 2.09° West

Our skin is a complex, adaptable organ that responds to our health and the world around us including the climate and weather. Differences in the weather between summer and winter and winter and spring can bring about changes in the appearance and texture of our skin, leading to dryness, itch, irritation and even skin disease. We can anticipate and prevent this by being aware of our environment, at home or abroad, and being ready to vary how we care for our skin.

We often turn to the weather when describing the place we call home because it affects almost everything we do, in what we choose to wear, the food we eat, our willingness to go outdoors and ultimately whether or not we can survive. In the UK, we live on an island in the Northern Hemisphere with a temperate maritime climate. The angle of the sun and tilt of the Earth mean that we receive less warmth from sunshine, while westerly winds from the Atlantic bring depressions with bands of cloud and rain, to which Scotland is particularly vulnerable[1,2,3]. Our combination of latitude and climate mean that nothing stands still - we are always passing through seasons, creating a rich variety of colours and things to do in the landscape around us.

❖

Our skin is a complex, adaptable organ that responds to our health and the world around us

❖

Cold weather, which we see a reasonable amount of in the UK, can make even the happiest of skins become dry, red and cracked. Wind, the cold and use of central heating all encourage our skin to dry out and partly as a result, eczema seems to be more common in colder, wetter, less sunny areas of the world[4,5]. The impact of cold weather extends to what we eat, as the warming snacks and comfort foods encouraged by long, dark winters may affect the balance of nutrients in our skin barrier and our antioxidant defence systems. Our pattern of occasional, unreliable sunlight at northern latitudes has an impact on our vitamin D levels and our risks of sun damage and skin cancer, and weeks of cloud, wind, rain and sleet can be draining for the body and mind.

❖

Good skincare repairs and prevents the damage caused by our environment

❖

But cold weather can be fresh and invigorating. Ice and frost bring sparkle to dull streets, and smoking log fires and cosy knits warm us from the inside out. Such a feeling of contentment can be gained from watching rain battering against a window from inside beside a roaring fire, or snuggling under a blanket while a frustrated wind howls around the rafters.

The lifestyle terms hygge and coorie, used to describe cosily comfortable recharging self-care, do not originate from cold, northern areas of the world by accident. The ideas behind these terms have probably been part of human resilience in the face of cold, harsh conditions for thousands of years. Skincare has a natural place in this, as good skincare repairs and prevents the damage caused by our environment, but also has relaxing, stress-relieving, emotional benefits and can be a useful habit to hoard away to stay feeling healthy in the cold.

This book explores how our skin responds to cold weather and the climate variables associated with northern latitudes. It discusses lifestyle habits that we adopt and the impact these have on our skin, and considers things we can do to stay looking and feeling our best in the cold. I hope it helps you to shine brightly on the dullest of days.

References

1. The British Geographer. Climate of the British Isles [Internet]. United Kingdom: The British Geographer; [updated unknown; cited 2018 June]. Available from: http://thebritishgeographer.weebly.com/the-climate-of-the-british-isles.html
2. The Royal Meteorological Society. UK Climate [Internet]. United Kingdom: The Royal Meteorological Society, MetLink; [updated unknown; cited 2018 June]. Available from: http://www.metlink.org/secondary/key-stage-4/uk-climate/
3. The Met Office. UK regional climates [Internet]. United Kingdom: The Met Office; [updated 2016 January; cited 2018 June]. Available from: https://www.metoffice.gov.uk/climate/uk/regional-climates
4. Silverberg JI, Hanifin J, Simpson EL. Climatic factors are associated with childhood eczema prevalence in the United States. J Invest Dermatol. 2013;133:1752-1759
5. Engebretsen KA, Johansen JD, Kezic S, Linneberg A, Thyssen JP. The effect of environmental humidity and temperature on skin barrier function and dermatitis. J Eur Acad Dermatol Venereol. 2016;30(2):223-249

1. Layer Up: An Introduction to the Science of Skincare

1. Layer Up:
An Introduction to the Science of Skincare

Our skin is one of the largest organs in our body. Most of our organs are protected, nestled under layers of tissue and bone, but our skin covers us like a shell or wrapping and is exposed to attacks from wind, water, sun and snow, bugs, beasties and whatever else it comes into contact with. Our skin protects us, warns us if we are near anything hot, cold or sharp, bends and stretches any way we fancy, stops us from losing water, generates enough vitamin D to give our bones strength and heals itself as and when required.

The reasons for and solutions to skin problems are deeply rooted in science. This chapter is an introduction to the structure and function of our skin, its resources for defending itself and its processes for repairing damage.

The reasons for and solutions to skin problems are deeply rooted in science.

Skin Structure

Looking down at our skin we just see a flat expanse of tissue, but if skin is cut into and looked at under a microscope it has three main layers, a bit like a cake or sandwich, as shown in the illustration below. The top, outermost layer is called the epidermis, beneath that is the dermis and then the deepest layer is the subcutaneous fat.

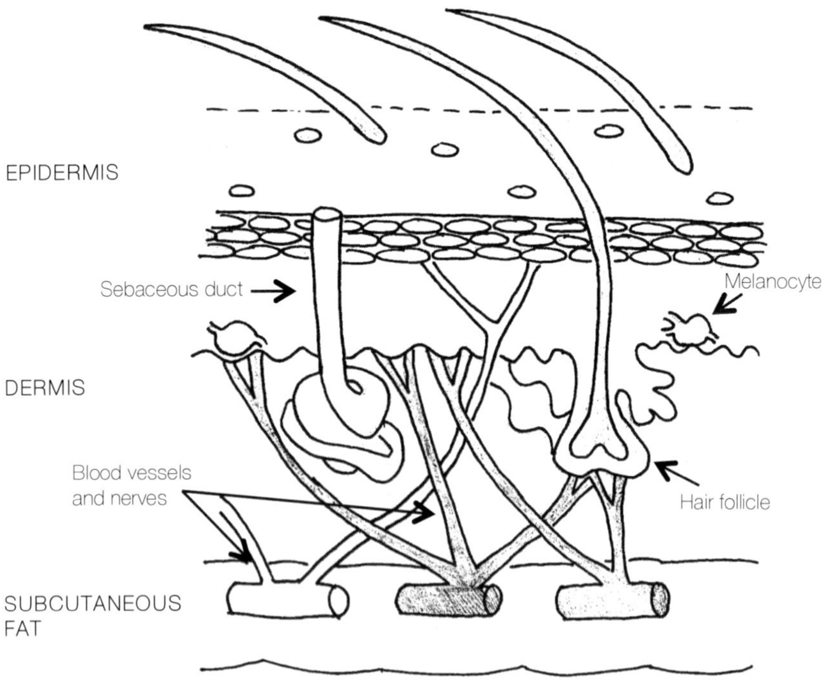

The Epidermis and the Skin Barrier

The epidermis itself consists of four microscopic layers: the stratum corneum at the surface, the stratum granulosum below, the stratum spinosum below that and the stratum basale at the bottom, neighbouring the dermis. The epidermis, particularly its top layer, the stratum corneum, is a barrier between the outside world and the inside of our bodies. It prevents harmful substances from entering, controls how much water passes out of our skin into the air and protects us from the effects of the sun's radiation.

Like other organs, skin is built from a variety of different cells. The epidermis contains several cell types, including keratinocyte cells, filled with keratin filaments, which act like bricks to give structure to the epidermis; melanocyte cells, which produce a brown pigment called melanin to provide protection from sunlight; Merkel cells, which are associated with nerves and the sensation of touch; and Langerhans cells, which communicate with the immune system to protect us from infection.

Keratins are the most important family of proteins in the epidermis. Keratin filaments form a structural network spanning the epidermis, both inside keratinocytes and outside in the space between cells. They attach to desmosomes (joins between cells) anchoring cells in place.

The epidermis varies in thickness depending on the area of the body. It is much thicker on the soles of our feet than it is on our eyelids, but on average it is 0.1mm thick, roughly as thick as a sheet of paper.

The Brick Wall

Many dermatologists describe the epidermal skin barrier as being like a brick wall[1]. Keratin filled keratinocyte cells in the stratum corneum act like bricks in the wall, giving it structure. The keratinocyte cell bricks are embedded in a lipid enriched matrix substance, which acts like mortar between the bricks, controlling the entry and exit of water. The lipid molecules here include ceramides, cholesterol and fatty acids arranged in an organised formation called the lipid bilayer. This lipid bilayer is our main defence against evaporation[1].

Healthy skin is like a new brick wall

Healthy skin is like a new brick wall with no cracks for water or bugs to leak out of or slip through. But if skin is irritated, broken or eczematous then it is as if the brick wall is crumbling, gaps are appearing and skin can dry out or become infected more easily.

Moisturisers can help repair damage to the skin barrier by providing a protective coating over the cracks and replacing, or encouraging the skin to produce, important molecules that allow the barrier to regain its strength.

Waterproofing

Our bodies need to be able to retain water or we will simply evaporate, and the epidermis itself needs to contain just the right amount of water in order to stay soft, supple and free from scaling.

The structure and components of the skin barrier in the stratum corneum control the balance of water in the skin[2]. Keratin, keratinocytes and the lipid bilayer are key players in this, as is a substance called natural moisturising factor (NMF), which is a collection of molecules that help to attract water into the epidermis and keep it there. Aquaporin water channels are involved in the transport of water and moisturizing molecules through the epidermis[3]. Problems with the skin barrier can be related to the structure and function of the lipid bilayer, to low levels of supporting substances like NMF or to problems with supporting structures like aquaporins.

❖
The skin barrier controls the balance of water in the skin
❖

Lipids

The volume of lipids in the skin seems to drop in winter, which will contribute to skin drying out[1]. In contrast, ultraviolet (UV) light in sunlight seems to increase levels of lipids in the epidermis[1]. Ceramides make up about half of the lipids in the stratum corneum, cholesterol makes up about a quarter and fatty acids make up about a sixth[3]. Ceramides are often mentioned in skincare articles and product adverts in relation to the ingredients of moisturisers. Ceramides prevent loss of water from the epidermis and the skin barrier performs its functions better when it contains higher levels of them[3]. Lips have a thin stratum corneum with lower levels of ceramides, which will contribute to a higher rate of water loss and skin that can easily become dry and cracked[4,5].

Natural Moisturising Factor

Natural moisturising factor (NMF) is a collection of water attracting substances. Half of it is amino acids, including pyrrolidone carboxylic acid (PCA). The other half includes lactate, urea and electrolytes[2,3]. These molecules attract water into the stratum corneum from the atmosphere and the deeper skin layers. Keratinocytes that contain most natural moisturizing factor hold on to more water, while lower levels are linked to skin dryness, scaling, flaking and cracking[3,6]. Our levels of PCA reduce as we get older and can also be lost from the skin by washing with soap[3].

Aquaporins
Aquaporin channels transport water and moisturizing molecules like glycerin between cells. Aquaporin-3 is the most common aquaporin in the epidermis and lives in the cell membrane of keratinocytes. The number of aquaporin-3 channels in skin reduces with age and with years of excessive sun exposure[3].

The Acid Mantle

The pH level of our skin is important in maintaining normal levels of healthy bacteria (the normal microflora), the function of the skin barrier, the creation and organization of epidermal lipids and the loss of cells from the stratum corneum through desquamation[7,8]. Skin should have a fairly acidic pH of 4-6, sustained by a thin film on our skin surface known as the acid mantle[7,8]. This film is created from a mixture of skin moisture, sebum and lactic acid from sweat and is influenced by the area of the body it covers, our age, genetics and ethnicity[8]. Alkaline skincare products, which can increase the skin's pH to above the normal range, have more potential to irritate and affect the skin barrier than pH neutral ones[8,9]. Soap is one of the most common alkaline skin products.

Desquamation

Desquamation is the process by which old cells are sloughed off from the top of the stratum corneum. This is necessary in order to prevent the build up of scale on the skin surface and the function of the skin barrier is affected if it goes wrong. Desquamation is dependent on water balance in the stratum corneum, as the enzymes needed to break the joins (desmosomes) between cells need water in order to function properly[2].

Responses to Sunlight

Dermatologists have a bit of a paradoxical approach towards sunlight and the UVA and UVB radiation it contains. We advise against too much sun exposure because of the risks of sun damage and skin cancer, but we also use UV light to treat many skin conditions.

> ❖
> Everyone should try and avoid sunburn, which is associated with skin cancer as a result of DNA damage
> ❖

UVB, and to a lesser extent UVA rays cause damage to cell DNA through the creation of pyrimidine dimers – lesions in DNA that can lead to problems in its function and accuracy of replication if they are not repaired or bypassed[10,11]. Inflammation and oxidative damage occur as UVB and UVA rays trigger keratinocytes to release cytokine messenger molecules and other inflammatory mediators, resulting in the production of reactive oxygen species (ROS) and free radicals like singlet oxygen and hydrogen peroxide, which damage cells and DNA[10,12,13]. Everyone should try and avoid sunburn, which is associated with substantially greater risk of skin cancer as a result of this DNA damage[10].

Sunlight can have good effects on the epidermis, for example it increases the volume of epidermal lipids; it triggers adaptations within the skin such as thickening of the epidermis and production of melanin production that protect its cells from sunlight (known as photoadaptation); it triggers production of vitamin D; it reduces the activity of the immune system in the skin, which can have unwanted effects but allows it to be used in treatment of some skin diseases; and it raises levels of a hormone called beta-endorphin in our bodies, creating a feeling of wellbeing[1,10,13,14].

❖
Intermittent sun exposure has implications for the health of our skin and for our health in general
❖

Not getting enough sun exposure (without burning) has been related to health risks including risks of several cancers, high blood pressure and diseases of the heart, lungs and blood vessels, diabetes, the metabolic syndrome, obesity, Alzheimer's disease, multiple sclerosis, rheumatoid arthritis, psoriasis, non-alcoholic liver disease and eye problems[10]. Some benefits of sun exposure can be related to the vitamin D produced in response to UVB rays but many others are thought to be independent of this, so taking a vitamin D supplement is not an entirely adequate substitute for natural sunlight[10]. The key of course is balance between sun exposure and sun protection, but where the balance lies will vary between each of us depending on characteristics like our skin colour, our age, our ability to go outdoors and where in the world we live.

Living at northern latitude means that we spend much of our lives with our skin covered in layers of clothing, living indoors out of the sun, or buried in dark winter months with no exposure to UVB, which is only present in sunlight between late spring and early autumn[15,16]. Because of this, our skin can be surprised and unprepared when it is eventually exposed to natural sunlight, which often comes in short sharp bursts during a few weeks of local good weather or during a sunny holiday abroad, in contrast to the constant exposure of people living in sunnier climates. This intermittent pattern of sun exposure has implications for the health of our skin and for our health in general.

Defences Against UV Damage

Protection against sunburn and skin cancer within our skin is thought to come partly from photoadaptation (the greater production of melanin and thickening of the epidermis that occurs in response to sunlight) and possibly the creation of higher levels of the active form of vitamin D[10]. Also, the inflammation and oxidative damage caused by UV light alerts damage response pathways in the skin, including processes for repairing DNA, processes to stop damaged cells from growing or producing new versions of themselves, processes encouraging damaged cells to die rather than linger and cause problems, and processes to reduce inflammation[10,13].

Melanin is a pigment that can absorb UV rays, helping to prevent damage to cell DNA. It is produced by melanocytes in the stratum basale (the bottom layer of the epidermis) and packaged into melanosomes. Melanosomes leave the melanocytes and are taken in by nearby keratinocytes, where they settle themselves above the DNA-containing cell nucleus. During exposure to sunlight, melanin is oxidized by UVA, while UVB encourages melanocytes to create more melanin[12,13].

We all have roughly the same number of melanocytes. The difference in colour between genetically light and dark skins is down to the activity level of our melanocytes and the type and volume of melanin that they produce. Human skin tones vary considerably and in research studies or medical notes these differences may be described using the Fitzpatrick Phototype Scale, created in 1975 by Dr Thomas Fitzpatrick.

Fitzpatrick Skin Phototypes

Type	Pigmentation	Response to Sunlight
Type 1	Pale skin, blue eyes, fair hair	Always burns, never tans
Type 2	Fair skin, blue eyes	Usually burns, tans little
Type 3	Fair to light brown skin	Tans well but may burn
Type 4	Light brown skin	Tans well, burns little
Type 5	Brown skin	Tans easily, rarely burns
Type 6	Dark brown to black skin	Never burns

Production of Vitamin D

At northern latitudes most of our vitamin D3 is made in our skin in response to the UVB rays in summer sunlight. We receive UVB in sunshine between roughly April and September, which acts on 7-dehydrocholesterol in the epidermis to produce vitamin D3[15,16]. Our diet provides small amounts of vitamin D, with vitamin D3 found in oily fish and egg yolk and vitamin D2 in plants and mushrooms. Vitamin D2 is the form used in food fortification and both vitamins D3 and D2 can be taken as supplements[17].

Vitamins D3 and D2 are inactive inside the body and need to be converted into the main circulating form, calcidiol (25OHD), by enzymes in the liver and then to the active form, calcitriol (1,25D), by enzymes in the kidneys. Calcitriol travels around the body binding to vitamin D receptors to produce its effects. There is a vitamin D receptor in nearly every tissue and thousands spread throughout the genome, controlling hundreds of genes[18]. The molecule commonly measured on blood tests to determine our vitamin D status is calcidiol (25OHD).

Calcitriol has a major role in maintaining our calcium levels, which have to be kept within a narrow range. Our body can increase calcium through several pathways, including absorbing more calcium across the gut, reducing losses at the kidney and releasing it from stores in our bones. Calcitriol is involved in triggering these processes, but calcium can also be released from bone stores independently of calcitriol. If we don't have enough vitamin D2 or D3 to make calcitriol, then a vital part of these processes is lost.

With no increase in gut or kidney absorption, the body will instead keep on pulling calcium from our bones, leading to them softening and the eventual development of rickets or osteomalacia[19].

The Dermis

Below the epidermis, the dermis is the skin's engine house. It contains blood vessels to provide the epidermis with vital nutrition and help maintain a stable body temperature, nerves to communicate sensation, follicles and sebaceous ducts to produce hair and sebum and sweat glands again to help maintain our body temperature.

The dermis consists of a gel-like substance made from glycosaminoglycans, including hyaluronic acid, held together by a matrix of collagen fibres to provide strength and organization and elastin fibres to provide resilience and flexibility[12].

The cells of the dermis include fibroblasts, which produce collagen

The cells of the dermis include fibroblasts, which produce collagen; macrophages, which consume invading substances like bacteria; dermal dendritic cells, also involved in the defence against infection; and histamine-containing mast cells. Other types of cell may visit the dermis depending on the presence of skin disease.

Sensation and Communication

The dermis is wired with nerve fibres that run alongside blood vessels. Nerves sense pressure, temperature and movement on the skin and relay these as impulses along fibres all the way to our brain. Signals from nerves also trigger responses from structures in the skin, such as release of sweat from sweat glands, elevation of hair fibres or widening of blood vessels leading to more flow of blood and flushing in the skin.

As well as communication through nerve fibres, cells in the dermis and the epidermis can communicate by releasing messenger molecules called cytokines, releasing hormones, or by the movement of message carrying cells from one place to another.

Temperature Control

Skin acts as an insulator, keeping us warm, and a radiator, letting extra heat out. It does this mainly by changes in blood flow (by widening or tightening blood vessels), but also by triggering sweating, as the evaporation of water from the skin removes heat[12]. If blood vessels widen, more warm blood flows into the skin, so more heat can be lost by radiation. This is why our skin goes pink or red when we exercise.

If we are exposed to very cold temperatures our body automatically begins to shiver to try and generate heat and we can feel the need to run or jump to make ourselves warm. In the cold, paradoxical widening of blood vessels at extremities like our fingers and toes can keep these areas warm and supplied with nutrients, but if we are really, really cold then these blood vessels tighten, reducing the flow of blood here in order to keep heat in the centre of the body, leading to tissue damage like frostbite[20]. Nerve impulses order muscles attached to small hairs in our skin to contract when we are cold, generating goosebumps and lifting the hair. In animals with significant body hair this response will help to retain body heat but it probably doesn't add much in humans. With less body hair, we have evolved 'artificial' ways of retaining and generating heat, like clothing, shelter and fire.

The Subcutaneous Fat

Our body fat exists under our skin in a continuous layer of cells underneath the dermis (the subcutaneous fat) and also scattered elsewhere in the body including around organs. Fat, also known as adipose tissue, usurped skin as being the largest organ in the body after its significant endocrine role was recognized at the start of the century. Fat cells can produce and release hormones and other signaling molecules to help regulate our appetite, blood pressure, glucose levels, immune system and reproductive ability[21]. Fat cells also provide insulation, an energy store and a soft and spongy layer to protect our internal organs from injury.

Skin Immunity

As well as being a physical barrier against water loss and the entry of bugs and beasties, our skin is a front line for our clever immune system in the fight against infection. Keratinocytes, matrix cells, messenger molecules including cytokines and immune cells like Langerhans and dermal dendritic cells all chatter amongst each other to control the reaction of the immune system to any invading organisms and to keep the skin environment in a happy equilibrium. The many tiny bugs and beasties that call our skin their home also make contact with our immune system and agree terms by which they can live side by side as a sustainable community[22].

❖
Skin is a front line for our clever immune system in the fight against infection
❖

For as long as this balance continues, our skin remains a happy, healthy place. But skin problems and inflammation can develop if invading microorganisms are felt to be too dangerous to ignore, or if the body's own immune cells begin to fight amongst themselves. Immune reactions leading to eradication of an infection are entirely wonderful, this is just what the process intends to achieve. But sometimes our body reacts too strongly to an infection, or sometimes it reacts to no infection at all, and in these situations medical treatment might be needed to coax the behavior of the immune system back to its happy equilibrium. Rosacea is one condition discussed in this book that is caused at least in part by an imbalance in the skin's immune system.

The process of reducing the activity of the immune system is called immunosuppression. Several skin treatments work by suppressing the activity of the skin's immune system. UV light produces some of its helpful effects in treating skin diseases in this way. But while this can be helpful in treating psoriasis and other skin diseases, the immune system is also involved in recognizing and fixing cell damage, so suppressing it for long periods of time may contribute to the chance of cancerous changes developing[12]. UV treatment for skin disease is given carefully with measured doses to ensure as far as possible that the benefit of its use outweighs any risks.

Antioxidants

Antioxidants are often referred to in relation to skincare and our diet but it can be unclear what they are and why we need them. Our cells need to generate energy to power everything from exercise to repairing sunburn and fighting off bacteria. Much of our energy is generated through a series of chemical reactions known together as oxidative phosphorylation, which takes place in microscopic cellular structures called mitochondria.

> ❖
> Oxidative phosphorylation is vital but produces potentially harmful reactive oxygen species and free radicals
> ❖

Oxidative phosphorylation is a vital process but it produces potentially harmful molecules and substances like reactive oxygen species (ROS), including free radicals, like hydroxyl radicals and superoxide, singlet oxygen and non-radical oxidants like hydrogen peroxide and nitric oxide[23,24]. A free radical is an unstable, highly reactive electron that detaches from a molecule during a chemical reaction and should reattach almost straight away. If left to themselves, they can cause mayhem by damaging cells and DNA.

Antioxidants include vitamins, pro-vitamins, enzymes, minerals, molecules and compounds, including vitamins A, C and E, glutathione, coenzyme Q10, zinc, copper and selenium, and catalase, peroxidase and superoxide dismutase. They act a bit like an organised cleaning service, travelling in the wake of oxidative phosphorylation and cleaning up its mess by attaching to free radicals and joining with or deactivating other ROS, limiting the damage inflicted on our cells, proteins and DNA.

A state called oxidative stress occurs when chemical reactions increase and the ROS and free radicals they produce overwhelm the ability of our antioxidants to mop them up. This may happen when our immune system reacts to invasion by bugs, or when we are exposed to chemicals or radiation, including UV light. If our antioxidant processes fail and oxidative stress persists then damage to proteins, lipids and DNA can eventually lead to cell death, replication of damaged genes and the development of cancers. There is almost no area of human disease in which oxidative stress has not been implicated and as we get older our antioxidants reduce in number and activity, making oxidative stress more likely to occur[23,24].

Interestingly, our antioxidant mechanisms are able to adapt to the stresses we put them under, becoming more effective. Taking part in exercise or restricting our intake of glucose or calories increases the activity of our mitochondria, increasing oxidative phosphorylation and production of ROS, and our antioxidant defence mechanisms adapt to respond more effectively, becoming more resistant to stress[25].

Reducing Inflammation

Inflammation is the process behind our skin becoming red, irritated, itchy, stingy, sore or swollen. Inflammation can appear for many, many reasons including in response to the stresses of cold weather on the skin. Similar cells and messenger molecules to those involved in skin immunity are largely involved in driving inflammation. Inflammation attracts the immune system, and the cells and messengers involved in the immune system can cause inflammation while trying to fix the issue that set it all off. The process can be triggered by some external insult or injury to the skin or by a disagreement within the skin itself.

❖
Many basic moisturisers help to reduce inflammation
❖

There are cells and messengers in the skin that try to act as a brake, controlling the amount of immune activity and inflammation occurring in an area. These restraining mechanisms are very effective as our skin is constantly bombarded with foreign objects, species and harmful rays that may threaten to throw it off balance. But in some cases the control processes do break down, leading to inflammation appearing and escalating[22].

I'm struggling to think of a skin treatment that doesn't exert some of its effect by reducing inflammation. Even many basic moisturisers help to reduce inflammation by restoring the skin barrier and rehydrating the epidermis. More sophisticated treatments for skin disease reduce inflammation by refereeing the messenger molecules in the skin, stopping pro-inflammatory messengers from causing trouble, allowing the curbing mechanisms to regain control and a calm equilibrium to return. Many of the elements discussed later in this book, either included as part of skin creams or eaten as part of our diet, may help to reduce the inflammation caused by cold weather.

DNA Repair

UV light is a significant threat to the DNA in our skin cells. UV is absorbed into skin cell DNA with the production of pyrimidine dimer photoproducts - lesions that form in DNA[10,11]. These lesions can be repaired by a process called nucleotide excision repair, in which the offending segment is chopped out, or they can be bypassed during the cell replication process. Failure to properly repair or avoid these lesions can start the chain of events leading to skin cancer[11].

Stem Cells

Our bodies change considerably throughout our lives. As time goes by, our cells die and are replaced with new ones, over and over again. This renewal process happens even without any injury or disease causing damage to tissue, but of course when damage does occur we also need to be able to make new cells in order to fully recover.

The process of cell renewal is driven by stem cells. These are unspecialized cells with the ability to divide and produce new baby cells of a given tissue. Stem cells are exciting in terms of medical research because of this potential to create new cells from scratch. Multipotent stem cells can create several cell types limited to those found within a particular tissue or organ, while pluripotent stem cells can potentially produce any cell that the human body needs[26].

The skin has two sets of stem cells, epidermal progenitor cells, found at the basal layer of the epidermis, which give rise to the cells of the epidermis, and hair follicle stem cells, found in an area of the follicle known as the bulge[26]. Hair follicle stem cells are thought to be multipotent in nature. They can create several structures found in the skin and help to repair the epidermis after injury[27]. Some creams claim to contain stem cells, sometimes stem cells from plants, but more research is needed to identify the molecular benefit these would have on human skin[28].

To renew the epidermis, epidermal progenitor cells at the stratum basale divide and produce new, fragile keratinocytes that move up through the epidermal layers, maturing as they go, until they reach the stratum corneum, where the old top layer is constantly being sloughed off through desquamation. This process may speed up if there is skin disease or inflammation but it usually takes about 4 weeks, meaning that you have a new top layer of skin from one month to the next[12].

Summary

- The reasons for and solutions to skin problems are deeply rooted in science.
- The epidermis contains our skin barrier, which regulates water balance, controls the entry of microbes and molecules and provides protection from injury, infection, chemicals and radiation.
- The dermis supports the epidermis in terms of nutrition and structural reinforcement. It also helps regulate body temperature and communicate sensation.
- Our immune system, antioxidant and anti-inflammatory processes, DNA repair mechanisms and stem cells all work together to keep our skin intact, healthy and in a happy equilibrium.
- Imbalance in these processes can lead to skin problems.

References

1. Jungersted JM, Hellgren LI, Jemec GB, Agner T. Lipids and skin barrier function – a clinical perspective. Contact Dermatitis. 2008;58(5):255-262
2. Kraft JN, Lynde CW. Moisturizers: what they are and a practical approach to product selection. Skin Therapy Lett. 2005;10(5):1-8
3. Draelos ZD. Modern moisturizer myths, misconceptions and truths. Cutis. 2013;91(6):308-314
4. Tamura E, Ishikawa J, Naoe A, Yamamoto T. The roughness of lip skin is related to the ceramide profile in the stratum corneum. Int J Cosmet Sci. 2016;38(6):615-621
5. Isoda K, Nakamura T, Yoshida K, Tamura E, Atsuta N, Ishida K, Takagi Y, Mizutani H. The efficacy of a lip balm containing pseudo-ceramide on the dry lips of sensitive skin-conscious subjects. J Cosmet Dermatol. 2018;17(1):84-89
6. Levin J, Miller R. A guide to the ingredients and potential benefits of over-the-counter cleansers and moisturizers for rosacea patients. J Clin Aesthet Dermatol. 2011;4(8):31-49
7. Proksch E. pH in nature, humans and skin. J Dermatol. 2018;45(9):1044-1052
8. Ali SM, Yosipovitch G. Skin pH: from basic science to basic skin care. Acta Derm Venereol. 2013;93(3):261-267
9. Draelos Z, Hornby S, Walters RM, Appa Y. Hydrophobically modified polymers can minimize skin irritation potential caused by surfactant-based cleansers. J Cosmet Dermatol. 2013;12(4):314-321
10. Hoel DG, Berwick M, de Gruijl FR, Holick MF. The risks and benefits of sun exposure 2016. Dermatoendocrinol. 2016;8(1):e1248325
11. Cleaver JE, Crowley E. UV damage, DNA repair and skin carcinogenesis. Front Biosci. 2002;1(7):d1024-1043
12. Bolognia JL, Jorizzo JL, Schaffer JV. Dermatology. 3rd ed. United States of America: Elsevier Saunders; 2012
13. D'Orazio J, Jarrett S, Amaro-Ortiz A, Scott T. UV radiation and the skin. Int J Mol Sci. 2013;14(6):12222-12248
14. Bulat V, Situm M, Dediol I, Ljubicic I, Bradic L. The mechanisms of action of phototherapy in the treatment of most common dermatoses. Coll Antropol. 2011;35(2):147-51 (abs only)
15. Lips P, van Schoor NM, de Jongh RT. Diet, sun and

lifestyle as determinants of vitamin D status. Ann N Y Acad Sci. 2014;1317:92-98
16. Scientific Advisory Committee on Nutrition (SACN). Vitamin D and Health. 2016 July [cited 2019 June] Available online from:https://assets.publishing.service.gov.uk/government/uploads/system/uploads/attachment_data/file/537616/SACN_vitamin_D_and_health_report.pdf
17. British Nutrition Foundation. Nutrients, food and ingredients. [Internet]. United Kingdom: British Nutrition Foundation; [updated unknown; cited 2019 June]. Available from: https://www.nutrition.org.uk/nutritionscience/nutrients-food-and-ingredients.html
18. Bikle DD. Vitamin D metabolism, mechanism of action, and clinical applications. Chem Biol. 2014;21(3):319-329
19. Laird E, Ward M, McSorley E, Strain JJ, Wallace J. Vitamin D and bone health: potential mechanisms. Nutrients. 2010;2(7):693-724
20. Daanen HA. Finger cold-induced vasodilatation: a review. Eur J Appl Physiol. 2003;89(5):411-426
21. Coelho M, Oliveira T, Fernandes R. Biochemistry of adipose tissue: an endocrine organ. Arch Med Sci. 2013;9(2):191-200
22. Pasparakis M, Haase I, Nestle FO. Mechanisms regulating skin immunity and inflammation. Nat Rev Immunol. 2014;14(5): 289-301
23. Addor FA. Antioxidants in dermatology. An Bras Dermatol. 2017;92(3):356-362
24. Lenaz G. Role of mitochondria in oxidative stress and ageing. Biochim Biophys Acta. 1998;1366(1-2):53-67
25. Ristow M, Schmeisser S. Extending life span by increasing oxidative stress. Free Radic Biol Med. 2011;51(2):327-336
26. Blanpain C, Fuchs E. Epidermal stem cells of the skin. Annu Rev Cell Dev Biol. 2006;22:339-73
27. Mistriotis P, Andreadis ST. Hair follicle: a novel source of multipotent stem cells for tissue engineering and regenerative medicine. Tissue Eng Part B Rev. 2013;19(4):265-278
28. Lee CM. Fifty years of research and development of cosmeceuticals: a contemporary review. J Cosmet Dermatol. 2016;15(4):527-539

2. Making Powder Sparkle:

Dry Skin in the Cold

2. Making Powder Sparkle:
Dry Skin in the Cold

As the weather turns from mild late summer into chilly golden autumn, temperature sensing nerves in our skin remind us that the season for bare legs is over, we need to untangle those scarves from the back of the cupboard and hunt down our missing gloves. Our skin loses its summer tan just as the shops fill with coats, hats and optimistic Christmas decorations.

Our skin can often dry out in the cold. Skin that is usually healthy may just feel more rough and itchy, but skin that already has a tendency towards dry conditions like eczema or psoriasis can be more likely to flare with the change in seasons. How dry our skin becomes depends on the climate and environment we are exposed to, the integrity of our skin barrier, our age, genetics and general health, and what we choose to do about it.

❖

Many of us find that our skin dries out over winter

❖

Skin dryness can be the main cause of symptoms like itch, stinging, cracking and irritation. Therapeutic (rather than heavily fragranced or cosmetic) moisturisers are part of the antidote to dry skin but these important skin treatments are often undervalued by non-dermatological medical professionals and under-recognised by patients. People quite often come to dermatology clinics using moisturisers that are at best unhelpful for their particular skin problem, or even no moisturiser at all.

Soap became easily available in the early 1900's and along with regular washing it has improved personal hygiene, reduced skin infections and changed the course of infectious diseases. But, because of the way that cleansers interact with the skin and components of the skin barrier, there is a delicate balance between skin hygiene and skin barrier damage, which will be tested even more during cold weather.

This chapter explores the impact weather has on our skin barrier and how to pick basic skincare products that offer the best protection from the elements.

The Skin Barrier and Skin Hydration

As discussed in Chapter 1, the stratum corneum at the outermost layer of the epidermis is the main structure controlling the entry and exit of water from our skin, and also requires just enough water itself in order to function properly[1]. Its regulatory role is down to its organised layer of lipid molecules - the lipid bilayer - made from ceramides, fatty acids and cholesterol. Molecules and structures like natural moisturising factor and aquaporin water channels are involved in keeping the skin hydrated.

Problems with the integrity of the skin barrier can lead to dryness, itch, stinging and irritation and be the basis of skin conditions including eczema. Problems can be caused by issues with the lipid bilayer, low levels of its supporting substances, or disruption in the pH balance at the skin surface. A normal, soft, supple stratum corneum is thought to contain 20-35% water, with both too little and too much water affecting its ability to function as an effective barrier[1,2,3].

❖
Problems with the integrity of the skin barrier can lead to dry skin
❖

One important water-dependent process in the stratum corneum is desquamation, in which old cells are constantly being sloughed away from the skin surface so new ones can replace them, keeping the skin barrier fresh and effective. New skin cells are formed from stem cells at the stratum basale and migrate up through the epidermis. As the new cells reach the stratum corneum, water-dependent enzymes break desmosomes linking the cells in the old top layer, allowing them to be lost[1]. The epidermal cell layers continuously renew themselves in this way roughly each month[4].

When the water level in the stratum corneum is low, these enzymes do not work properly and keratinocytes gather on the skin surface in a traffic jam, producing scaling[1]. Some of the ingredients of moisturisers try to smoothen or break up these crowds of cells and may be used in combination with scrubs, brushes and other exfoliants to encourage renewal of the skin surface.

The acid mantle has an important role in maintaining the normal function of the skin barrier. People with a tendency towards eczema and skin irritation seem to have higher (more alkaline) pH values at the surface of both affected and unaffected skin compared to people with healthy skin, suggesting that disruption in skin surface pH is part of the disease process[5]. Using moisturisers, cleansers and skincare products that encourage an acidic environment, as opposed to alkaline preparations, should help to maintain skin's pH at acceptable levels, supporting the skin barrier and microflora[5,6].

Weather and the Skin

Skin is a living organ that responds to our surroundings, adapting to protect the DNA in its cells, keeping it free from disease and performing as it should[7]. But despite this ability skin can still be affected by environmental changes including cold weather, leading to problems like dryness and itch.

In general, the condition of our skin can deteriorate in cold weather[8]. This may be partly related to changes in barometric air pressure, which can encourage movement of water up through the epidermis to evaporate from the stratum corneum[9]. This is combined with a seasonal drop in the volume of lipids and natural moisturising factor in the stratum corneum, which will also encourage loss of water, skin fragility and irritation[9].

❖

Skin is a living organ that responds to our surroundings

❖

Several scientific studies have looked at the effect individual weather elements like humidity, temperature and ultraviolet (UV) light have on our skin, but obviously in real life these elements are met in combination, with all the variety that the Northern Hemisphere can throw at us. There are not many research studies looking at the effect of real life weather conditions or exposure to these over long time periods[7].

While there is a group of eczema sufferers who find that high temperatures and lots of sunshine worsen their skin, research results indicate that, in general, dry skin conditions are more common at northern latitudes, particularly over winter[9,10]. Skin is likely to dry out in environments that are cold, dry and windy[9,11]. Of the research studies examining combinations of weather conditions, one from the United States involving 91,642 children found that eczema – a skin condition caused by leakiness of the skin barrier leading to skin dryness and inflammation - was more common in children living in areas with more rainfall and central heating use. The weather combination of high humidity, high rainfall and little strong sunlight was associated with more eczema[12].

Lets consider the evidence surrounding the main weather elements individually.

Humidity

Low humidity (dry air) is often blamed for skin problems. During cold weather we often crank up our heating systems, which dry out the air around us. Sudden exposure to a dry environment causes dermal mast cells to release histamine, increases other inflammatory mediators, affects the skin barrier, reduces the water content of the stratum corneum and increases skin roughness, dryness, itch and inflammation[7,9,13,14]. Dry air makes it harder for the desmosome links between cells to break, reducing desquamation and leading to a build up of scale[9,13].

Relative humidity indicates how much water vapour is in the air as a percentage of the maximum amount the air could hold[9,13]. 90% relative humidity is a very moist environment and 10% is very dry. Humans prefer relative humidity to remain between roughly 25 and 60%, with 45% being the most comfortable[13,15].

Warm air is able to hold more water than cold air can[9]. When cold air is heated, by central heating for example, its ability to hold water increases and so the relative humidity of the environment drops (it becomes drier)[9]. We spend between 60 and 90% of our lives indoors, so dry air here really has a great opportunity to affect our skin[16]. Room sensors can be bought to indicate humidity levels.

A Guide to Relative Humidity (RH):	
Dry sauna	5-30%[17]
Steam room	100%[17]
Aberdeen, Scotland	Annual average 85%[18]
Madrid	Annual average 61%[18]
Timbuktu, Mali	Annual average 28%[18]

Exposure to humidity can be short, from minutes in a sauna to hours or days in a workplace; or longer, to days or years, depending on where you live and what you do. The results from one scientific study indicate that exposure to a dry environment can make fine wrinkles become more visible, particularly at the face and around the eyes where there is a thinner stratum corneum[19]. Another study, which looked at the effects of short 30 minute exposures, suggested that moving from an environment with 70% relative humidity to a drier one with 40% relative humidity significantly increased the area of fine wrinkles at the corner of the eye, and reduced the stretchiness and moisture content of the stratum corneum here[20]. This result suggests that, at least in terms of wrinkles and skin moisture, it may be best to visit the sauna before the steam room, rather than the other way around!

But, a balance exists between too much and too little humidity, as damp, humid environments are not good news either. In another population study, involving 5595 children enrolled from every region of the United States, high humidity was associated with poorly controlled eczema, with a 10% increase in the chance of poorly controlled symptoms with every 10% increase in humidity. It was felt that this might be related to the effects of sweat, which can irritate eczematous skin and is less able to evaporate in humid environments[10].

Increasing relative humidity and using appropriate skincare can improve skin symptoms. Because of close links between humidity and temperature, lowering the thermostat heating temperature by a few degrees can increase room humidity, and spending more time outdoors also reduces our exposure to dry air[12,14]. Decent occlusive skincare can prevent and treat the skin barrier damage, dryness, roughness and fine lines from dry environments[7,9].

Increasing relative humidity and using appropriate skincare can improve skin symptoms related to a dry environment.

Temperature

Human beings are sensitive to even small temperature changes, to the extent that temperature has been described in architectural literature as being the most significant determining factor in finding comfort in a space[15]. The temperature of our environment affects our skin temperature directly by convection and also by triggering responses from nerves and blood vessels in the skin[7]. The sensation of itch can become more intense when the surrounding temperature increases or decreases[9,10].

❖

There is less water in the stratum corneum at colder temperatures[9]

❖

In warm temperatures, sweating and greater blood flow to the skin mean that more water is lost across the stratum corneum[8]. As the air gets colder, loss of water from the epidermis reduces rapidly, as sweating stops and blood flow to the skin reduces to conserve heat[7]. Despite the reduction in water loss, there is still less water in the epidermis at colder temperatures and in general, cold air has detrimental effects on the skin barrier[9].

Eczema is more common in colder parts of the world[21]. A research study observing 3000 children from around the world found that eczema became increasingly common at more northern latitudes and areas with lower outdoor temperatures[21]. This may be partly due to differences in lifestyle at northern, colder areas, differences in hours and intensity of sunlight, and higher use of central heating, with its impact on relative humidity[14,21].

Wind

Water is constantly evaporating from the surface of our skin. Wind contributes to the loss of water by blowing away airborne water particles at the skin's surface, reducing the humidity in that particular area[11]. The reduction in humidity allows more water to evaporate from the skin, so, the windier it is across the surface of your skin, the more water is encouraged to evaporate.

❖

The windier it is across the surface of your skin, the more water is encouraged to evaporate

❖

Pollution

Free, unlimited access to clean air is a basic human need and right[22]. Air quality depends on the amount of pollutant emissions, physicochemical changes in the atmosphere and the movement of polluted air masses[23]. Ambient air pollution consists of several hundred different components including particulate matter (PM), ozone, nitrogen dioxide, sulphur dioxide and carbon monoxide. PM itself is a mixture of nitrates, sulphates, ammonium and carbons[22].

Apart from ozone, which reaches peak levels in the summer, levels of air pollution peak in the winter and drop over the summer[22,23]. This is because of a combination of factors including emissions from home furnaces, lower air temperatures creating smog events and more use of cars in colder weather. Also, during sunnier months some pollutants react with light, which can reduce their levels[23].

❖

Levels of air pollution peak in the winter and drop over the summer[23]

❖

Ambient PM, including PM10 (particles with an aerodynamic diameter of 10μm) causes skin problems by triggering oxidative stress. Along with solar radiation and cigarette smoke, pollution has been linked to speeding up the signs of ageing. Highly reactive nanosize particles, particularly those from traffic, are considered to be the most damaging[24]. As well as the well-reported associations with lung problems, pollution may cause skin disease to flare up more often[22,25].

Moisturisers: an Antidote to Dry Skin

Moisturisers are made from active ingredients, which improve the condition of the skin, and excipients such as preservatives and emulsifiers, which come together in a delivery system. The most effective moisturisers aim to improve the hydration of the stratum corneum, maintain the integrity of the skin barrier, maintain the skin's naturally acidic pH, reduce the symptoms of dry skin such as itch and tightness and may improve wellbeing by contributing to a healthy appearance. Moisturisers are the most common component of skincare products, the most commonly prescribed products in dermatology and probably one of the most ancient treatments known to man[3,26,27].

❖
No pill or potion can produce the same improvement in dry skin, with the same safety and tolerance, as a decent moisturiser
❖

'Moisturiser' is a marketing term, and products may not actually add any extra moisture or water to the skin[1,27]. Many products may produce their effect by trying to stop the water that is currently in the skin from evaporating. There are a huge number of moisturisers available and it is important to separate those moisturisers that are marketed solely for their fragrance or brand from therapeutic moisturisers, whose primary aim is to improve dry skin[28]. Many people view moisturisers as a cosmetic product rather than an active treatment, which can undermine belief in their medical benefits[3,27]. A lack of faith, plus practical difficulties like the time it takes to apply creams and their texture, can mean that some people are less willing to use them regularly. This is to the detriment of their skincare, as at present no pill or potion can produce the same improvement in dry skin, with the same safety and tolerance, as a decent moisturiser.

The Best Moisturiser

Fittingly, the moisturiser market is saturated, with products made by dermatological companies, cosmetic companies, small businesses and entrepreneurial individuals, plus those mass made for celebrities, hotel chains, home ware companies and supermarkets. Every dermatologist will have been asked "what is the best moisturiser?", but one single product is not going to be better than all others, for every individual, at every time of year, in every situation. The problems within the skin leading to dryness vary between people, depending on genetics, presence of skin disease and products being used. The same product may not even suit the same person all the time as the condition of our skin varies between seasons, with our environment and with our general health.

❖
Cost and packaging can be a poor guide when looking for an effective moisturiser
❖

Products with a range of active ingredients have more of a chance of working well as they are more likely to fix whatever factor or combination of factors had caused the skin to become dry[28]. Cost and packaging on the other hand can be a poor guide to a good product. Knowledge about effectiveness, stability and interactions between ingredients is important in order to be able to recommend a suitable product, and dermatologists can explain the differences between marketing claims, anecdotal evidence and proven clinical effects[3,28]. There is quite a bit of trial and error in finding a moisturiser that works for you, which can be frustrating. In choosing between the therapeutic products, personal preference of delivery system and how the product looks feels and smells can be the ultimate deciding factor. In several ways, the best therapeutic moisturiser for you is the one you prefer, as then it will be used[3].

Delivery Systems

Most moisturisers are two phase emulsions, with one substance containing droplets of another, resulting in preparations described as lotions, creams or ointments[3,29]. Thicker products are often better suited to treating dry skin, with creams seen as more effective than lotions and ointments the most effective, but with increasing thickness comes stickiness and a shiny appearance.

While many commonly available lotions are just too thin to treat dry skin, some therapeutic products have been made incorporating active ingredients within this more convenient delivery vehicle. These lighter products can be easier to use regularly as they spread more easily on the skin and soak in quickly, making them more likely to be used rather than sit on the shelf[28].

Lotions

Lotions tend to be thinner in consistency with lower oil content than creams, making them convenient to use for daytime skincare, particularly on the face[3]. They are easy to spread across the skin so can be a useful vehicle for suncream, or products whose primary function is to deliver fragrance. Non-comedogenic face lotions are suitable for acne-prone or oily skin as their chosen ingredients plus thin texture reduces the chance of pores clogging up. Lotions historically have only been suitable for use on skin that is not particularly dry, but as above, increasingly effective moisturising lotions are being produced[28]. Typical constituents of a lotion include propylene glycol, water or mineral oil[1].

Creams

Creams are the most common type of delivery system used for therapeutic moisturisers and are generally made with heavier lipids including petrolatum and lanolin[1,3]. The ratio between oil and water is important, with products with higher oil content being more greasy[3]. Formulations for different skin types and body sites are based on adjustment of the oil : water ratio and which occlusive and emollient ingredients are used[1].

Ointments

These are a single-phase system rather than an emulsion and thick in texture. They become easier to apply when warmed between the hands. Once applied, they sit on top of the skin and provide a raincoat-like protective coating, reducing loss of water from the skin and preventing further damage by external irritants or the environment. Ointments may not always be practical to use on dry skin in the daytime, particularly the hands, but they can be a very effective boosting treatment under cotton gloves or clothing overnight. Nappy preparations and lip balms often come in ointment form.

Ingredients

The qualities of a moisturiser are determined by its ingredients[30]. Ingredients can be active, aiming to improve the skin barrier, or excipients, which stabilise the formulation, maintain the blend of ingredients, and prevent it from going off. Extra ingredients intended to be rejuvenating, antioxidant or anti-inflammatory are also often added.

To make a good moisturiser, ingredients should be included in physiologically relevant concentrations, be part of a stable formulation, be suitable for sensitive skin (described as hypoallergenic, non-sensitising, fragrance free and/or non-comedogenic), create a product that is slightly acidic in pH and make the product affordable, long lasting and quickly absorbed[1].

❖
The qualities of a moisturiser are determined by its ingredients
❖

Active ingredients include components of natural moisturising factor (such as lactates and urea), ceramides and barrier lipids, or ingredients that can increase the production of these in the skin[28]. Active ingredients can be divided into three main groups: emollients, humectants and occlusives, and may produce their effects by coating the surface of the stratum corneum or by penetrating the skin barrier and working within it. Emollients soften and smoothen keratinocytes at the skin surface, improving the appearance and texture. This immediate effect makes it more likely that the

product will be used again. Humectant ingredients attract water from their surroundings, while occlusives coat the stratum corneum, providing a waterproof protective film. An ingredient may have the properties of more than one group, for example silicones like dimethicone are both emollient and occlusive in action.
Nearly all moisturisers contain a combination of ingredients as many work better when applied together, for example glycerin's ability to supplement natural moisturising factor improves when it is delivered in combination with emollients, and lipids are more effective when combined with glycerin[29]. Together, glycerin, dimethicone and petrolatum form the basis of many skincare products[2,26]. Also, different ingredients work in different ways and choosing a product that addresses several possible factors contributing to skin dryness will lead to an improvement in symptoms in most cases[28].

Some important ingredients and excipients can have unwanted effects on the skin. Emulsifiers are needed to keep the ingredients blended, but can weaken the skin barrier. Preservatives are needed to prevent mouldiness, but can cause contact allergy in some people. Some humectants, occlusives and emollients can also have unwanted effects, for example petrolatum may sometimes delay skin barrier recovery; propylene glycol may produce irritant dermatitis in sensitive patients; alpha hydroxy acids (AHAs) like lactic acid can cause stinging and irritation; and lanolin can cause allergic or irritant reactions[30]. Whether an ingredient causes irritation to any given person can also depend on the amount of it contained in the product[30].

Occlusives

Examples[1,2,26,27,29]:	Petrolatum, paraffin, mineral oil (also known as liquid paraffin and paraffinum liquidum), cetyl alcohol, silicon derivatives such as cyclo- and dimethicone, squalene, lecithin, propylene glycol, cholesterol, carnauba, beeswax, lanolin, stearin.

Occlusives coat the stratum corneum, reducing loss of water across the epidermis and can be particularly effective when applied to damp skin[1,26]. They protect the skin from irritation and contribute to the components of the matrix between keratinocytes, allowing barrier repair[1,28]. Occlusives are often combined with humectants as humectants draw water into the stratum corneum and the occlusives help keep it there[3,26,28].

❖
Occlusives coat the stratum corneum, reducing loss of water.
❖

A heavily occlusive product may be less pleasing to look at and less practical because of its greasiness. Some can have a particular smell and some can become contact allergens[1]. Some occlusive agents like mineral oil can be comedogenic, blocking pores and leading to worsening of acne and acne-like conditions. If you do suffer from acne or an acne-like rash, or you find that some moisturisers give you more spots, look for formulations that state they are non-comedogenic and/or non-acnegenic.

The two compounds most commonly identified as occlusives are liquid paraffin and petrolatum[1,2,27]. Petrolatum is one of the most commonly used occlusives and active ingredients in skincare and has been used in skincare since 1872[2,26,31]. It creates a protective film and helps to repair the skin barrier by increasing production of lipids in the epidermis[2,26,32]. It reduces the appearance of fine lines by also acting as an emollient, filling the spaces between desquamating keratinocytes and acts as an exfoliant by encouraging these cells to slough off[2,26]. It is hypoallergenic, non-comedogenic and non-acnegenic[2].

Petroleum jelly is the most effective occlusive. In as little concentration as 5% it reduces loss of water across the epidermis by more than 98%[1,27]. It is followed by lanolin, mineral oil and silicones, which reduce epidermal water loss by 20-30%[1,29].

Lanolin is the most commonly used animal wax[3]. It is a natural product that can't be synthesised and comes from the sebaceous secretions of sheep[27,31]. It contains cholesterol, an essential component of the epidermal lipid bilayer. It is possible to develop contact allergy to lanolin but ultrapure medical grade lanolin's have been developed that aim to reduce the chance of this[31].

Dimethicone is a common ingredient of oil-free moisturisers. Dimethicone belongs to the family of silicones, originating from silica found in sand, quartz and granites[3]. It is hypoallergenic, non-comedogenic and non-acnegenicand reduces water loss across the epidermis but does not stop it completely[2,26]. Dimethicone is less greasy than petrolatum and more permeable to water vapour, but will not reduce water loss if the skin barrier is compromised, while petrolatum will[2,26]. The two agents are often used together for their relative benefits[2]. Dimethicone can also act as an emollient.

Humectants

Examples[1,2,26,27,28,29]:	Glycerin, urea, lactic acid, sodium, potassium and ammonium lactate, sodium pyrrolidone carboxylic acid (PCA), hyaluronic acid, carnitine, amino acids including arginine hydrochloride, serine, alanine, histidine, lysine, threonine, citrulline, glutamic acid; sodium chloride, glycogen, mannitol, sucrose, sorbitol, gelatin, honey, panthenol, propylene glycol.

Humectants are absorbed into the stratum corneum, making products pleasant to use and practical for use in the daytime. They attract water to the stratum corneum from the atmosphere above and the deeper epidermis and dermis below. Their actions can actually lead to more water evaporating from the stratum corneum, especially if the skin barrier is compromised, so to avoid that they must be combined in formulations with occlusives[1,2,26].

Glycerin (also called glycerol) is a commonly used and very effective humectant[1,2,26,27]. Glycerin and petrolatum (an occlusive) both continue to benefit the skin long after they are applied. In glycerin's case this is thought to happen by an effect on aquaporin water channels[2,26]. Glycerin can also encourage the maturation of any fragile, immature keratinocytes that may reach the stratum corneum at exposed body sites during winter months, increasing the strength of the skin barrier[1,29].

Urea has been used for decades to treat dry skin and can increase the volume of important skin proteins like filaggrin (involved in maintaining adequate levels of natural moisturising factor), the volume of enzymes needed to generate lipids, and molecules involved in immunity such as cathelicidin and defensin[33].

Humectants including glycerin, urea, propylene glycol and alpha hydroxy acids (AHAs, especially lactic acid) can also act as exfoliants, promoting desquamation[1,3,29,34].

❖
Humectants attract water to the stratum corneum.
❖

Components of natural moisturising factor like PCA, urea and lactic acid support the function of the skin barrier. Lactic acid can stimulate production of ceramides in the skin and its acidic pH should complement the skin's naturally acidic environment, important for maintaining normal microflora and barrier function[1,5].

Emollients

Examples[1,3,26,27,28,29]:	Essential saturated fatty acids (stearic, lauric acids), unsaturated and polyunsaturated fatty acids (linoleic, oleic, linolenic acids) found in natural oils such as coconut oil, jojoba oil and wool fat. Glycol stearate, soy sterol, sunflower seed oil, waxes, cetearyl alcohol, mineral oil, lanolin, dimethicone, cyclomethicone, isopropyl myristate, isopropyl palmitate, isostearyl alcohol, glyceryl stearate, squalane, propylene glycol, petrolatum.

❖
Emollients soften and smoothen keratinocytes, improving the appearance, texture and flexibility of the skin
❖

Emollients are mainly lipids and oils that soften and smoothen keratinocytes and fill in the crevices between cells, improving the appearance, texture and flexibility of the skin[1,26]. The action of emollients can appear to reduce fine lines and these quick, visible and textural improvements can encourage people to continue using the products[1,26].

Squalene is an ingredient of sebum and one of the most common lipids produced by human skin cells[27]. Production reduces after about the age of 30 however, and lower levels thereafter contribute to skin becoming drier with increasing age[27]. Squalene has antioxidant action and can protect the skin from sun damage[27,35]. Squalane is a saturated form of squalene and is less susceptible to oxidation, so is more commonly used as an ingredient in moisturisers[27,35]. Squalane is non-comedogenic, antibacterial and safe for sensitive skin[27].

Preservatives and Emulsifying Agents

Examples[3,30,36]:	
Preservatives:	Sodium benzoate, potassium sorbate, sorbic acid, phenoxyethanol, parabens, DMDM hydantoin, benzyl alcohol.
Emulsifiers:	Polysorbates, potassium cetyl sulphate, laureth-4, lecithin, glyceryl stearate, cetyl alcohol, ceteareth 20, polyethelene glycol (PEG), triethanolamine.

Preservatives keep products fresh and free from harmful bacteria. The ideal preservative should act against a wide range of bacteria, be safe and stable and not affect the other ingredients of the moisturiser[3,27]. Usually a combination of substances is used to achieve this[3,27]. Antioxidants including vitamins E and C and chelating agents like citric acid can assume the role of preservative within a formulation[3].

Unfortunately, preservatives can become contact allergens. As problems with preservatives are found, the amounts of them in products may be reduced or they fall out of use to be replaced with alternatives. Methylisothiazolinone was a commonly used preservative but the European Union placed a ban on its use in leave-on products like moisturisers in 2016 as a result of its tendency to cause contact allergy[37]. Formaldehyde releasing compounds like DMDM hydantoin and quaternium-15, often used in cosmetics, are also common contact allergens and the use of these has also changed significantly since they were first introduced[38].

Emulsifiers are also necessary ingredients that bring potential issues. They are needed to keep the water-loving and water-hating parts of a moisturiser in a blended formulation and are similar to the substances used as surfactants in cleansers. They can be anionic (negatively charged), nonionic (no charge) or cationic (positively charged). Anionic compounds including stearic and palmitic acids have the most potential to annoy the skin barrier[30]. Nonionic emulsifiers including cholesterol, polyethylene glycol (PEG), cetearyl alcohol, and stearyl alcohol are usually less irritating, but PEG is susceptible to oxidation and formation of reactive oxygen species (ROS)[3,30]. Cationic emulsifiers like benzylkonium chloride, quaternium 15,19,23, stearalkonium chloride can have good and bad effects on the skin[30]. Benzylkonium chloride is often included in antimicrobial preparations for its antimicrobial benefits, but it can irritate the skin and remove healthy skin microflora[36].

Cosmeceuticals and Additional Ingredients

Examples:	Vitamins, antioxidants, peptides, botanical extracts, marine extracts, alpha- and beta-hydroxy acids (AHAs and BHAs), minerals, growth factors.

It is useful to remember that the pleasant effects of many products - like softening the skin and reducing visible fine lines - are because of the moisturising ingredients they contain[26]. The range of additional ingredients in moisturisers and cosmetics is becoming more and more diverse, with many plant extract (botanical), antioxidant and cosmeceutical compounds appearing, many of which aim to improve the signs of ageing and sun damage[3,28].

Cosmeceuticals are substances that may have good effects on the skin's structure or function but not enough of an effect as to be classed as a drug. They are a booming industry but the term 'cosmeceutical' is a marketing term rather than a legal one. While prescription medicines must be scientifically demonstrated to improve or cure a condition, and are governed by the Medicines and Healthcare Products Regulatory Agency (the MHRA), the regulation currently surrounding cosmeceuticals in the UK and Europe in 2019 does not require demonstration of clinical effect, but does ensure safety, accurate labelling, and that the products are not tested on animals[39].

The real-life effectiveness of any ingredient will depend on its concentration and stability within the product, as well as its ability to penetrate the stratum corneum if it aims to replenish the nutrients and skin substances usually found deeper in the skin. Given the competition in the industry, the effectiveness of any product may only be known by the manufacturing company.

Some of the more common ingredients added to moisturisers, cosmetics and other body care products are described below.

Aloe Vera, or Aloe Barbadensis Leaf Extract

Aloe Vera has been used for many hundreds of years as a botanical ingredient in skincare. It contains many molecules that may have active effects on the skin, including vitamins, enzymes, minerals and fatty acids, and may have anti-inflammatory and antibacterial effects. While some small studies and reviews have been performed, at present no clinical trials have been completed to determine its effectiveness, though the length of time that is has been used in skincare would suggest that it could produce good effects[40].

Alpha-Hydroxy Acids (AHAs)

AHAs include glycolic, citric and lactic acids. Formulations with low concentrations, like those that can be bought for use at home, aim to exfoliate the skin surface, reducing scaling and roughness[41]. They can encourage moisture to stay in the skin by acting as humectants and can increase the production of ceramides by keratinocytes[1,3,5].

Allantoin (comfrey root)

Allantoin is a botanical ingredient. It has been used as a medicinal plant for centuries, particularly to reduce pain and inflammation, and may also improve wound healing. Its anti-inflammatory effect has been found to equal that of diclofenac and it has also been used for muscle and joint complaints[42]. It may improve skin hydration and reduce roughness and scaling by encouraging desquamation[3].

Caffeine

Caffeine can be included to reduce inflammation and may constrict facial veins to prevent flushing. It has a dehydrating effect that is commonly used in anti-cellulite creams, where it may last for around 24 hours[43].

Cannabinoids

Cannabinoids are a diverse group of compounds. Some are naturally produced in the bodies of humans and animals, some are derived from plants, and some are made in laboratories. They may be able to improve a variety of skin complaints, including dry skin, inflammation, itch and even skin cancers, but more research is needed to determine their mechanisms of action, effectiveness and safety[44].

Chamomile

Chamomile is a botanical ingredient that has been used for centuries to treat skin conditions. Clinical studies (and the duration of its use) support its supposed antimicrobial, antiallergenic, anti-inflammatory, antioxidant and pain relieving properties. Its active ingredients include isoprenoids, flavonoids and oligosaccharides[45].

Co-Enzyme Q10

Co-enzyme Q10 is a molecule that assists with the production of energy in cell mitochondria. In its reduced state (known as ubiquinol) it is a potent antioxidant, providing protection from free radical damage, with an effect roughly equivalent to that of the active form of vitamin E (alpha tocopherol)[46,47]. It may protect against sun damage, resist breakdown of collagen and reduce inflammation produced in response to sunlight[46]. It may also reduce the depth of wrinkles[48].

Despite the positive results from laboratory tests, in real life co-enzyme Q10 needs to penetrate the stratum corneum in order to produce these effects. There has been some evidence that it can penetrate and other data suggesting it may have considerable difficulty, but as very low concentrations are needed, enough may penetrate from most formulations in order for it to have an effect[46,48]. Overall it seems possible that co-enzyme Q10 may have favourable effects on the skin, and as levels in our cells reduce with age and lack of exercise it is a reasonable molecule to include in skincare[46,47].

Growth Factors

Growth factors are signalling molecules found within the body, usually involved in inflammation and wound healing. They include epidermal growth factor (EGF), platelet derived growth factor (PDGF) and granulocyte colony stimulating factor (GCSF). Their potential for skin rejuvenation is harnessed in preparations like platelet rich plasma (PRP), administered with the intent of rejuvenating and firming the skin by stimulating fibroblast cells to produce new collagen[41].

Hyaluronic acid (HA)

HA is a glycosaminoglycan and has the ability to attract water up to 1000 times its own volume. It is a key component of the dermis and very important in skin hydration. It also plays important roles in cell growth, membrane function and cell adhesion. Topical (applied to the skin) HA is available, but as it is a large molecule its ability to penetrate far into the skin can be questioned. It may act as a humectant to help hydrate the top of the stratum corneum. HA is the main component of several injectable cosmetic fillers[31].

Liquorice Extract

Liquorice extract is a botanical ingredient that contains flavonoids. It may reduce the production of melanin in the skin, which may help treat uneven pigmentation[41].

Polyphenols (Flavonoids, phenolic acids, anthocyanins)

The effects produced by botanicals in skincare mainly tend to be due to the phytochemicals and polyphenols they contain. These compounds have antioxidant activity, reacting with ROS and free radicals. Examples of these plant extracts include ginkgo biloba, ginseng, grape seed extract, green tea, lavender, rosemary, soy, aloe vera, allantoin and bakuchiol.

Polyphenols are a group of substances including flavonoids, tannins and phenolic acids, originally from plant based foods, that can also now appear in chemically modified or synthetic forms. Flavonoids include catechin, epicatechin and epigallocatechin-3-gallate, found in green tea, cocoa and grape seed extract. Polyphenols are described as anti-inflammatory, antioxidant, UV absorbing, and may help to prevent DNA damage, cell damage and cell death[27,49]. There is some clinical trial evidence to support this for oral and topical use[49].

Grape seed extract (vitis vinifera) is rich in powerful antioxidant compounds like proanthocyanidin and flavanoids and some experimental data suggests that grape seed proanthocyanadins (GSPs, containing anthocyanins) may protect against cancerous changes caused by sunlight, possibly by reducing the skin immunosuppression caused by UV[50,51]. Topical 2% grape seed extract has been shown to improve wound healing, possibly by triggering the release of growth factors and production of collagen, and may also have anti-inflammatory and antimicrobial properties[50].

Bakuchiol is a phenol compound isolated from an Indian plant called Psoralea corylifolia (babchi). It has been the subject of recent research and is believed to have anti-inflammatory, antioxidant and anti-acne effects, as well as possibly reducing the chance of cancerous change in the skin. It seems to target similar pathways to retinoids, and may produce effects comparable to retinol without the associated side effects[52].

Nicotinamide

Nicotinamide is the amide form of niacin (vitamin B3), also known as niacinamide. It is a synthetically produced water-soluble vitamin, and depending on its concentration it may be used to treat sun damage, itch and inflammation, uneven skin pigmentation, combat infection, reduce sebum production and improve dry skin[31]. It is a component of several pathways that work to resolve oxidative stress[41].

Nicotinamide is felt to be safe in a concentration up to 4%, though for treatment of uneven pigmentation concentrations up to 5% have been used. It increases the rate of ceramide synthesis and stabilises histamine containing mast cells, preventing them from bursting and releasing their contents. It is well tolerated and safe but there is little clinical data available to support its actual effectiveness[53].

Oatmeal (avena sativa)

Ground oatmeal has also been used in skincare for hundreds of years. It may have antioxidant and anti-inflammatory properties through the action of phenols[54]. Oat bran is a good source of B vitamins, vitamin E, fats and minerals[54]. It is reported to be soothing when added to baths as 10mls sprinkled directly under the tap, and may also be added to creams, cleansers, shampoos and shaving gels. There is limited scientific data on how effective it is, but the persistence of its use would suggest that it might produce good effects.

Omega 3- and 6- fatty acids

These polyunsaturated fatty acids may be included as emollient ingredients but their chemical structure means they are prone to oxidation. Topically applied omega-3 eicosapentaenoic acid (EPA) may protect the skin from the harmful and ageing effects of sunlight[55].

Polypodium Leucotomos

Polypodium leucomotos is a fern extract also known as Fernblock. It can be included in creams or taken as an oral supplement and has been described as anti-inflammatory, antioxidant and providing protection from sunlight by stopping production of ROS, preventing DNA damage, reducing inflammation, protecting internal antioxidant systems and reducing the volume of cell death[56,57]. This may reduce the chance of sunburn and help repair sun damage, protecting the skin from premature ageing as well as potentially the development of skin cancer[57]. It may help in the management of polymorphic light eruption, vitiligo, melasma, psoriasis and eczema as an add-on to existing treatments and has no reported side effects[56,57].

Teas: Black, White and Green

These are botanical ingredients containing polyphenolic compounds, which may provide protection from inflammation, oxidative stress and sun damage when applied onto the skin[45]. White tea seems to be a more potent antioxidant than Black tea[45]. The main active ingredient in Green tea is epigallocatechin-3-gallate (EGCG), which may protect the skin from inflammation, oxidation and sun damage. Green tea polyphenols may work with sunscreens to protect the skin, and may also have antibacterial properties[45].

Vitamin A, including retinoids (eg. retinol, retinaldehyde, tretinoin) and carotenoids (eg. beta-carotene, lycopene and lutein)
Vitamin A is essential for the normal structure and function of the skin and other body organs. Its antioxidant action traps and neutralises free radicals and protects cells from oxidative damage, which may protect against skin cancer[58]. The vitamin A family is known as retinoids, and carotenoids act as pro-vitamins that can be converted in the body into retinoids. Within the body, retinoic acid, primarily all-trans-retinoic acid, coordinates most of the functions of vitamin A. All-trans-retinoic acid (ATRA) has been available on prescription (known as tretinoin) since around 1962. All-trans-retinoic acid is produced as the end product of a retinoid cascade, and in terms of skincare, all over the counter, non-prescription retinoids (such as retinyl palmitate, retinol, retinaldehyde) must be converted to retinoic acid (all-trans-retinoic acid, ATRA) by enzymes in the skin in order to have an effect[59]. The cascade of conversion (simplified to present retinoids related to skincare ingredients) is shown below:

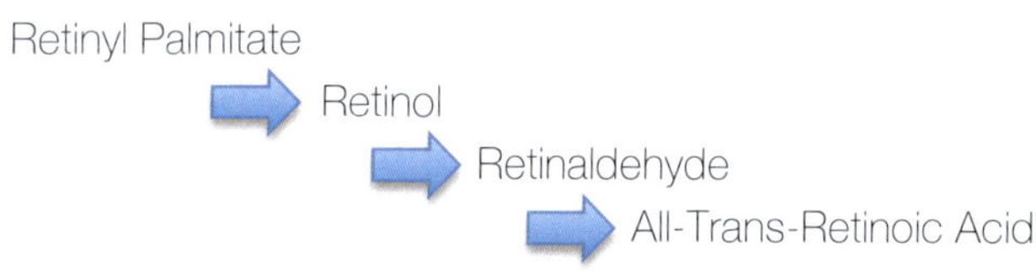

In theory, the closer a retinoid preparation is in the cascade to ATRA, the easier its conversion in the skin may be, and so it may be more effective. Carotenoids in topical preparations or the diet can be converted in the body to retinaldehyde (also known as retinal) and then to all-trans-retinoic acid. Retinol in topical preparations, supplements or diet can be converted to retinaldehyde, then all-trans-retinoic acid[59]. Topical retinol can influence dermal glycosaminoglycans and collagen, and may produce an improvement in skin wrinkling, skin signs of ageing and photoageing, but the vehicle used is crucial in its effectiveness, as retinol is unstable and easily degraded to inactive forms on exposure to light and air[60,61]. This is part of the reason that it is advised to apply these products at night-time.

Tretinoin (the prescription-only synthetic version of all-trans-retinoic acid) has been investigated the most in the topical treatment of photoageing[41]. Several months' treatment with 0.05% tretinoin cream improves the appearance of photo-aged human skin. It may also improve the appearance of chronologically aged skin, but there is less evidence for this and more studies are needed[60]. Tretinoin can cause skin irritation and redness and while the over the counter retinoids may be less effective, they tend to produce less side effects[41].

Carotenoids have strong antioxidant activity but their use has been limited by their intense red colour. Recent development of a mixture of colourless carotenoids avoids this problem, and is reported to provide the same level of antioxidant protection[46]. In laboratory testing, colourless carotenoids were seen to act with co-enzyme Q10 to reduce inflammation from exposure to sunlight, and they also protected it from degradation[46].

Vitamin C

Vitamin C is a water-soluble antioxidant often included as an anti-ageing and anti-inflammatory ingredient. Being water-soluble, its antioxidant activity occurs in the fluid inside and outside cells[58,62]. Vitamin C is needed for synthesis of collagen and for wound healing. Topical active vitamin C may enhance the production of collagen with possible anti-ageing benefits[63]. Vitamin C regenerates and works with vitamin E. It deactivates the free radicals produced by sunlight and may reduce the chance of sunburn, and exposure to sunlight and pollution reduces Vitamin C levels in the skin[63].

Active vitamin C is such an effective antioxidant that it is very unstable and easily oxidised on exposure to air, meaning that the shelf life of most formulations containing pure vitamin C is short[63]. If not properly formulated, topical preparations containing it may not work, or may be irritating to sensitive skin. To solve this problem, stable derivatives such as ascorbyl-6-palmitate and magnesium ascorbyl phosphate may be used, but there is the chance they may not be well absorbed or metabolised in the skin to the active form[63].

Vitamin E

As a lipid soluble antioxidant, vitamin E exerts antioxidant activity from within cell membranes, reacting directly with free radicals and deactivating ROS, protecting the membranes from oxidative damage[58,62]. Vitamin E is especially plentiful in the stratum corneum, delivered there by sebum, and may help to improve skin dryness particularly in people taking oral retinoids[31,63]. Like vitamin C, vitamin E levels are reduced by exposure to sunlight and pollution[63]. Topical vitamin E may reduce the wrinkles and pigmentation associated with sun damage[63].

Vitamins C and E stabilise each other and act together in cells to provide antioxidant protection with Vitamin E in cell membranes backed up by vitamin C in cell and matrix fluids, and vitamin C also regenerates vitamin E[58,63]. Vitamin E regenerates beta-carotene and protects vitamin A from oxidation[58]. Vitamin E may provide some protection from sunlight when combined with other antioxidants including vitamin C, but neither vitamin was demonstrated to provide sun protection when used by themselves[49,63].

Zinc

Zinc has been used as a medical treatment for centuries. It is essential for the regulation of lipid, protein and nucleic acid metabolism, for normal hormonal and reproductive function, for stable immune activity, for wound repair and for control of inflammatory processes. It also has antioxidant properties[64]. 6% of zinc in the body is held in the skin[64]. Skincare preparations such as zinc oxide, calamine and zinc pyrithione have been used for soothing properties, sun protection and in antidandruff shampoos, and zinc has been suggested for use in rosacea, eczema, psoriasis and sun damage, amongst other skin disorders[64].

Stinging

Stinging is an important side effect of moisturisers that can stop them from being applied regularly and can cause distress, particularly for young children. Stinging doesn't generally occur when products are used on unbroken skin but can happen when they are used on irritated or broken skin. There are several ingredients that are more associated with stinging when applied to irritated skin, including lactic acid, urea (20-40% of people affected with eczema develop stinging with urea), PCA (pyrrolidone carboxylic acid, part of natural moisturising factor), propylene glycol, and the preservatives benzoic acid and sorbic acid[3]. As a general rule, ointments are blander (contain less potentially irritating ingredients) than creams and so are less likely to sting when applied to irritated or broken skin.

Cleansers: be Sparkly Clean, not Squeaky Clean

Many of the skin oils, dirt and other unwanted substances that we aim to remove by washing tend not to dissolve in plain water (they are hydrophobic, or water-hating) and so are difficult to remove without cleansers, which contain ingredients called surfactants to remove them. There are many health and social benefits to be had from regular washing and use of cleansers, but harsh cleansers are one of the most common causes of damage to the skin barrier[65]. Choice of cleanser is very important for healthy skin, especially in colder weather.

> Harsh skin cleansers are one of the most common causes of damage to the skin barrier[59].

Surfactants

Cleansers can exist as bars, liquids, creams or waters, but all of them contain surfactants in order to produce their effect. Surfactants in skin cleansers are usually soaps or synthetic detergents (syndets). The choice of surfactant included in a cleanser has a major impact on how mild or harsh the formulation is[30].

Single molecules of surfactant (surfactant monomers) have one end that is able to dissolve in water and another that is able to dissolve in fat. This means they are attracted to the skin surface during washing, where oily substances meet water[66,67]. Surfactant monomers tend to assemble together into ring-like arrangements called micelles, with their water loving ends on the outside of the ring and their fat loving ends on the inside[66]. Micelle rings can then surround the oily skin dirt and sebum, dispersing these water-hating substances in water and allowing them to be rinsed off[66,67].

However, surfactants are not able to tell the difference between our unwanted skin dirt and our necessary skin barrier fats and proteins like ceramides, cholesterol and elements of natural moisturising factor[36,67]. Surfactant monomers and micelles can insert themselves into the lipid layers of our skin barrier and weaken it by disrupting the organization of lipids, dissolving enzymes and natural moisturising factor and affecting the maturation of keratinocytes and their ability to desquamate from the stratum corneum[66]. This damage leads to inflammation, irritation and oxidative stress, with skin that feels tight, red, itchy, rough and dry[36,66,67]. The 'squeaky clean' feeling achieved after washing with traditional soap is related to this skin barrier damage[36].

❖
Using cleansers containing gentle surfactants can reduce damage to the skin barrier
❖

Using cleansers containing gentle surfactants can reduce skin barrier damage, which is especially helpful for those of us whose skin barrier is already weakened by cold weather or skin problems[67]. While they might not be explained on the packaging, formulation characteristics like the surfactant surface charge, the size and shape of micelles and the overall pH of a product determines how gentle it is[66,67]. The terms ‘natural’ and ‘organic’ are often included on packaging, but these are used for marketing appeal and have no meaning in terms of the ability of the product to cleanse or irritate[36].

The surface charge of surfactant monomers and micelles may be negative (anionic), positive (cationic), positive and negative (amphoteric) or have no charge (nonionic). Examples and characteristics of these are shown in the table below.

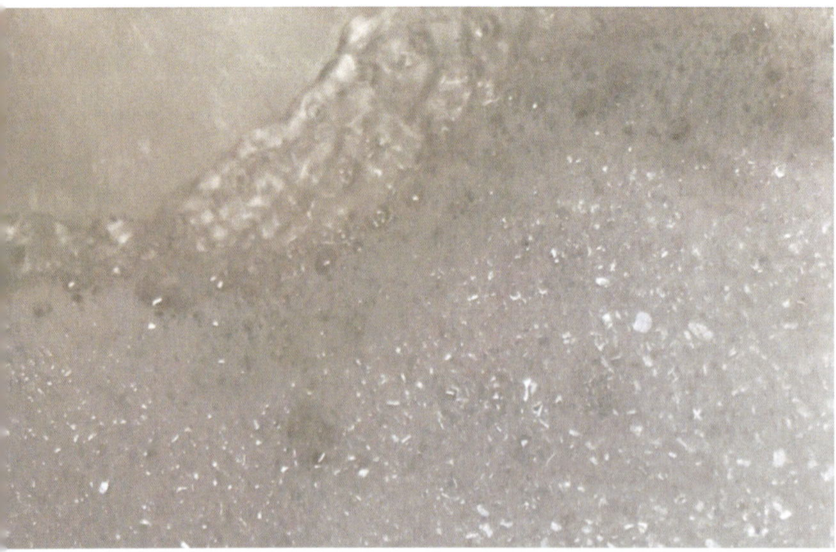

Surface Charge	Examples	Characteristics
Anionic	Soap Some syndets: sodium lauryl sulphate (SLS), sodium laureth sulphate (SLEs), sodium cocoyl isethionate[36,67]	Effective cleansers Foam well Potentially most irritating[30,68]
Amphoteric	Some syndets: Cocamidopropyl betaine[36,67]	Effective cleansers Foam Can irritate[67]
Nonionic	Some syndets: alkyl polyglucoside[36,67] Polysorbate 20	Gentle but effective cleansers Do not foam well[67]
Cationic	Benzylkonium chloride[36]	Antimicrobial Can irritate and remove normal skin flora[36]

Milder cleansers can be created by blending anionic surfactants with less irritating ones, but this may reduce their foaminess[36,66,67,68]. The ability of a product to foam is important commercially, as many people associate the presence of foam and lather with a product that cleans well[36].

The pH of cleansers can affect the naturally acidic environment of the skin surface. This acidic environment, known as the acid mantle, is important for maintaining our skin barrier and the normal skin microflora[5,6]. Alkaline cleansers tend to increase the skin's pH away from its comfort zone, so have more potential to irritate and harm the skin barrier than neutral ones[67]. Soaps tend to have an alkaline pH of 9-11 while syndets have a more neutral pH of 5.5 – 7, making cleansers with syndet ingredients a better choice for those of us with skin problems (and most people, really)[30,36]. Antimicrobial cleansers containing cationic surfactants also tend to have a slightly alkaline pH, around 7-10[5,30,36].

> ❖
> Alkaline cleansers like soap increase the skin's pH away from its acidic comfort zone[66]
> ❖

In order to limit skin barrier damage from surfactants, cleansers should be in contact with the skin for as short a time as possible before being rinsed off[36]. All surfactants have the potential to irritate, even mild ones, and surfactants that remain on the skin after cleansing may eventually lead to irritant contact dermatitis[36,66].

Polymer Technology

Adding polymers to cleansers can help to protect the skin[67]. For example, polyethylene oxide reduces the impact of surfactants on the skin barrier by interacting with micelles. This was the technology first used to make baby shampoo[66]. Hydrophobically modified polymers (HMPs) are a recent invention allowing products to cleanse gently at a low pH, avoiding irritation while still producing that important foam. HMPs attach themselves onto the water-hating ends of surfactant molecules resulting in a larger compound that is less able to enter the stratum corneum, meaning less barrier damage and irritation[36,66,67,68].

Cleansing Preparations

Surfactants are prepared and packaged into many different varieties of cleanser. While you can be pretty confident that a bar of soap bought in a soap shop will contain soap, it can be harder to know at a glance what is in many commercially available cleansers.

Soap

Soap is its own chemical entity, separate from syndets and other surfactants. Humans have been making soap for thousands of years using the process of saponification - mixing a fat with an alkali to produce a fatty acid salt with cleansing properties[66].

Soap is an anionic surfactant with a typical pH of 9-11[36]. Its chemical structure includes a carboxyl head group that loves to bind to proteins, which makes for excellent cleansing but causes damage to the skin barrier[30,36,65]. Soap contains chains of fatty acids, the length of which can influence the amount of damage it inflicts. Soaps with many short chains, which tend to come from plants, are less damaging than the longer chains, derived from animals[36].
Because of its alkaline nature, soap can increase the skin's pH leading to changes in microflora, epidermal

lipids and the skin barrier as a whole[6,36,67]. Damage can lead to swelling in the stratum corneum, allowing soapy surfactant molecules to penetrate deeper into the skin[30,36,65,67]. After rinsing, skin feels tight, dry and 'squeaky clean' because soap binding reduces the ability of keratinocytes and skin proteins to hold onto water[36]. Using soap-free cleansers can avoid these issues and help to reduce skin dryness, itch and irritation[65].

Bar Cleansers

Bar cleansers are often referred to as 'a bar of soap' but many are now made using syndets. A typical bar cleanser is a combination of the soap alkyl carboxylate and the syndet acyl isethionate. These are both anionic surfactants, with the soap's carboxyl group causing the most damage to the skin barrier[36]. Including additional compounds such as stearic acid has been used to produce milder, more moisturising cleansing bars[36].

Liquid Cleansers

Just as bar cleansers may contain syndets, liquid cleansers may contain soaps, with consequent higher pH levels[5]. In liquid cleansers, syndets and/or soaps are added to water, and moisturising ingredients like petrolatum, vegetable oils or shea butter may also be added[36]. At the start of washing, cleanser surfactants bind to oil-soluble dirt and prepare it for removal. Then, when water is added for rinsing, moisturising ingredients can be deposited on the skin in a thin film[36].

Facial Cleansers

Preparations like cleansing cold creams, cleansing milks, cleansing oils and cleansing balms are a good choice for cleaning dry skin, as while they can effectively remove cosmetics they don't remove as much moisturising sebum from the skin's surface[36]. They are wiped over the face and then wiped off with a tissue, rinsed, or left on. Preparations may include glycerin, cetyl alcohol or stearyl alcohol to leave behind a moisturising film[30,36]. Some of these preparations may have a tendency to block pores leading to spots (comedogenicity) in susceptible people.

Micellar waters effectively remove cosmetics and can be used on all skin types including dry, sensitive skin[36,65]. They are dilute cleansing solutions containing water and a very mild surfactant arranged in micelles. As in other cleansers, the micelles dissolve water-hating dirt in water by pulling it into their lipid-loving centres, then rinsing removes the particles from the skin[36]. Micellar waters contain mild, nonionic, non-foaming surfactants like polysorbate 20 and amphoteric surfactants similar to those in baby shampoo[36]. While many of these products state they can be left on the skin without rinsing, it is worthwhile considering the potential for any surfactant to cause irritation if in contact with the skin for long periods, particularly if you have especially sensitive skin. Rinse them off if you notice any irritation.

Summary

- Cold weather can lead to skin dryness, irritation and itch, related to the effects that low humidity, cold temperatures, wind and less sunlight have on the skin barrier.
- Increasing the relative humidity of your environment and using appropriate skincare can improve skin symptoms.
- The ingredients of a moisturiser determine its quality and character, while cost and packaging can be a poor guide.
- Harsh anionic and alkaline cleansers can damage the skin barrier leading to dryness, tightness and irritation. Using soap-free cleansers with gentle surfactants can help reduce barrier damage.
- Using a good moisturiser and a gentle cleanser everyday can help to protect the skin from the effects of cold weather.

References

1. Kraft JN, Lynde CW. Moisturizers: what they are and a practical approach to product selection. Skin Therapy Lett. 2005;10(5):1-8
2. Draelos ZD. Active agents in common skin care products. Plast Reconstr Surg. 2010;125(2):719-724
3. Loden M. The clinical benefit of moisturizers. J Eur Acad Dermatol Venereol. 2005;19(6):672-688
4. Bolognia JL, Jorizzo JL, Schaffer JV. Dermatology. 3rd ed. United States of America: Elsevier Saunders; 2012
5. Ali SM, Yosipovitch G. Skin pH: from basic science to basic skin care. Acta Derm Venereol. 2013;93(3):261-267
6. Proksch E. pH in nature, humans and skin. J Dermatol. 2018;45(9):1044-1052
7. Singh B, Maibach H. Climate and skin function: an overview. Skin Res Technol. 2013;19(3):207-212
8. Cravello B, Ferri A. Relationships between skin properties and environmental parameters. Skin Res Technol. 2008;14(2):180-186
9. Engebretsen KA, Johansen JD, Kezic S, Linneberg A, Thyssen JP. The effect of environmental humidity and temperature on skin barrier function and dermatitis. J Eur Acad Dermatol Venereol. 2016;30(2):223-249
10. Sargen MR, Hoffstad O, Margolis DJ. Warm, humid and high sun exposure climates are associated with poorly controlled eczema: PEER (paediatric eczema elective registry) cohort 2004-2012. J Invest Dermatol. 2014;134(1):51-57
11. Bolton N. Why does humidity and wind speed affect evaporation? [Internet]. United States: Sciencing.com; [updated 2018 March; cited 2019 June]. Available from: https://www.sciencing.com/humidity-wind-speed-affect-evaporation-12017079.html
12. Silverberg JI, Hanifin J, Simpson EL. Climatic factors are associated with childhood eczema prevalence in the United States. J Invest Dermatol. 2013;133:1752-1759
13. Goad N, Gawkrodger DJ. Ambient humidity and the skin: the impact of air humidity in healthy and diseased states. J Eur Acad Dermatol Venereol. 2016;30(8):1285-1294
14. Reinikainen LM, Jaakkola JJ. Significance of humidity and temperature on skin and upper airway symptoms. Indoor Air 2003;13(4):344-352
15. Boduch M, Fincher W. Standards of human comfort:

relative and absolute. Seminar Paper. University of Texas at Austin School of Architecture. Meadows Seminar in Sustainable Architecture; 2009. Available online from: https://www.soa.utexas.edu/sites/default/disk/preliminary/preliminary/1-Boduch_Fincher-standards_of_human_comfort.pdf
16. Jantunen M, Oliveira Fernandes E, Carrer P, Kephalopoulos S. Promoting actions for health indoor air (IAIAQ). European Commission Directorate General for Health and Consumers. 2011. Available online from: http://www.iaquk.org.uk/ESW/Files/env_iaiaq.pdf
17. Almost Heaven Saunas. Sauna vs. Steam room [Internet]. United States of America: almostheavensaunas.com; [updated 2013 March; cited 2019 June]. Available from: https://almostheaven.com/2013/03/steam-room-vs-sauna/
18. World Weather and Climate Information. Weather-and-climate.com [Internet]. The Netherlands: World Wide Travel Organisation; [unknown; cited 2019 Jun.]. Available from: https://weather-and-climate.com
19. Egawa M, Oguri M, Kuwahara T, Takahashi M. Effect of exposure of human skin to a dry environment. Skin Res Technol. 2002;8(4):212-218
20. Tsukahara K, Hotta M, Fujimura T, Haketa K, Kitahara T. Effect of room humidity on the formation of fine wrinkles in the facial skin of Japanese. Skin Res Technol. 2007;13(2):184-188
21. Weiland SK, Husing A, Strachan DP, Rzehak P, Pearce N. Climate and the prevalence of symptoms of asthma, allergic rhinitis, and atopic eczema in children. Occup Environ Med. 2004; 61(7):609-615
22. World Health Organisation (WHO). Health aspects of air pollution: results from the WHO project "systematic review of health aspects of air pollution in Europe". E83080 [internet]. 2004 June [cited 2019 June]. Available from: https://www.euro.who.int/_data/assets/pdf_file/0003/74730/E83080.pdf
23. Cichowicz R, Wielgosinski G, Fetter W. Dispersion of atmospheric air pollution in summer and winter season. Environ Monit Assess. 2017;189(12):605
24. Vierkotter A, Schikowski T, Ranft U, Sugiri D, Matsui M, Kramer U, Krutman J. Airborne particle exposure and

extrinsic skin ageing. J Invest Dermatol. 2010;130(12):2719-2726
25. Xu F, Yan S, Wu M, Li F, Xu X, Song W, Zhao J, Xu J, Kan H. Ambient ozone pollution as a risk factor for skin disorders. Br J Dermatol. 2011;165(1):224-225
26. Nolan K, Marmur E. Moisturizers: reality and the skin benefits. Dermatol Ther. 2012;25(3):229-233
27. Sethi A, Kaur T, Malhotra SK, Gambhir ML. Moisturizers: the slippery road. Indian J Dermatol. 2016;61(3):279-287
28. Draelos ZD. Modern moisturiser myths, misconceptions and truths. Cutis 2013;91:308-314
29. Rawlings AV, Canestrari DA, Dobkowski B. Moisturizer technology versus clinical performance. Dermatol Ther. 2004;17(1):49-56
30. Levin J, Miller R. A guide to the ingredients and potential benefits of over-the-counter cleansers and moisturizers for rosacea patients. J Clin Aesthet Dermatol. 2011;4(8):31-49
31. Baumann L, Saghari S, Weisberg E. Cosmetic dermatology: principles and practice. 2nd ed. United States of America: McGraw-Hill Companies Inc.; 2009
32. Grubauer G, Feingold KR, Elias PM. Relationship of epidermal lipogenesis to cutaneous barrier function. J Lipid Res. 1987;28(6):746-752
33. Grether-Beck S, Felsner I, Brendan H, Kohne Z, Majora M, Marini A, Jaenicke T, Rodriguez-Martin M, Trullas C, Hope M, Elias PM, Krutmann J. Urea uptake enhances barrier function and antimicrobial defence in humans by regulating epidermal gene expression. J Invest Dermatol. 2012;132(6):1561-1572
34. Madison KC. Barrier function of the skin: "la raison d'etre" of the epidermis. J Invest Dermatol. 2003;121(2):231-241
35. Kelly GS. Squalene and its potential clinical uses. Altern Med Rev. 1999;4(1):29-36
36. Draelos ZD. The science behind skin care: cleansers. J Cosmet Dermatol. 2018;17(1):8-14
37. Lundov MD, Opstrup MS, Johansen JD. Methylisothiazolinone contact allergy – growing epidemic. Contact Dermatitis. 2013;69(5):271-275
38. Fasth IM, Ulrich NH, Johansen JD. Ten-year trends in contact allergy to formaldehyde and formaldehyde-releasers. Contact Dermatitis 2018;79(5):263-269

39. The European Parliament and the Council of the European Union. Regulation (EC) no. 1223/2009 of the European Parliament and of the Council of 30 November 2009 on cosmetic products, L342/59 [Internet]. 2009 [cited 2019 June]. Available from: https://ec.europa.eu/health/sites/health/files/endocrine_disruptors/docs/cosmetic_1223_2009_regulation_en.pdf
40. Surjushe A, Vasani R, Saple DG. Aloe vera: a short review. Indian J Dermatol. 2008;53(4):163-166
41. Lee CM. Fifty years of research and development of cosmeceuticals: a contemporary review. J Cosmet Dermatol. 2016;15(4):527-539
42. Staiger C. Comfrey: a clinical overview. Phytother Res. 2012;26(10):1441-1448
43. Velasco MV, Tano CT, Machado-Santelli GM, Consiglieri VO, Kaneko TM, Baby AR. Effects of caffeine and siloxanetriol alginate caffeine, as anticellulite agents, on fatty tissue: histological evaluation. J Cosmet Dermatol. 2008;7(1):23-29
44. Eagelston LRM, Kalani NK, Patel R.R., Flaten HK, Dunnick CA, Dellavalle RPl. Cannabinoids in dermatology: a scoping review. Dermatol Online J. 2018;24(6). pii: 13030/qt7pn8c0sb
45. Stallings AF, Lupo MP. Practical uses of botanicals in skin care. J Clin Aesthet Dermatol. 2009;2(1):36-40
46. Fuller B, Smith D, Howerton A, Kern D. Anti-inflammatory effects of CoQ10 and colourless carotenoids. J Cosmet Dermatol. 2006;5(1):30-38
47. Beyer RE. The participation of coenzyme Q in free radical production and antioxidation. Free Radic Biol Med. 1990;8(6):545-565
48. Hoppe U, Bergemann J, Diembeck W, Ennen J, Gohla S, Harris I, Jacob J, Kielholz J, Mei W, Pollet D, Schachtschabel D, Sauermann G, Schreiner V, Stab F, Steckel F. Coenzyme Q10, a cutaneous antioxidant and energiser. Biofactors. 1999;9(2-4):371-378
49. Evans JA, Johnson EJ. The role of phytonutrients in skin health. Nutrients. 2010;2(8):903-928
50. Hemmati AA, Foroozan M, Houshmand G, Moosavi ZB, Bahadoram M, Maram NS. The topical effect of grape seed extract 2% cream on surgery wound healing. Glob J Health Sci. 2014;7(3):52-58
51. Katiyar SK. Grape seed proanthocyanidines and skin

cancer prevention: inhibition of oxidative stress and protection of the immune system. Mol Nutr Food Res. 2008;52(1):S71-76
52. Dhaliwal S, Rybak I, Ellis SR, Notay M, Trivedi M, Burnley W, Vaughn AR, Nguyen M, Reiter P, Bosanac S, Yan H, Foolad N, Sivamani RK. Prospective, randomised, double-blind assessment of topical bakuchiol and retinol for facial photoageing. Br J Dermatol. 2019;180(2):289-296
53. Wohlrab J, Kreft D. Niacinamide – mechanisms of action and its topical use in dermatology. Skin Pharmacol Physiol. 2014;27(6):311-315
54. Pazyar N, Yaghoobi R, Kazerouni A, Feily A. Oatmeal in dermatology: a brief review. Indian J Dermatol Venereol Leprol. 2012;78(2):142-145
55. Kim HH, Cho S, Lee S, Kim KH, Cho KH, Eun HC, Chung JH. Photoprotective and anti-skin-aging effects of eicosapentaenoic acid in human skin in vivo. J Lipid Res. 2006;47(5):921-930
56. Choudry SZ, Bhatia N, Ceilley R, Hougeir F, Lieberman R, Hamzavi I, Lim HW. Role of oral Polypodium leucomotos extract in dermatologic diseases: a review of the literature. Drugs Dermatol. 2014;13(2):148-153
57. Gonzalez S, Gilaberte Y, Philips N, Juarranz A. Fernblock, a nutriceutical with photoprotective properties and potential preventive agent for skin photoaging and photoinduced skin cancers. Int J Mol Sci. 2011;12(12):8466-8475
58. Rutkowski M, Grzegorczyk K. Adverse effects of antioxidative vitamins. Int J Occup Med Environ Health. 2012;25(2):105-121
59. Blomhoff R, Blomhoff HK. Overview of retinoid metabolism and function. J Neurobiol. 2006;66(7):606-630
60. Mukherjee S, Date A, Patravale V, Korting HC, Roeder A, Weindl G. Retinoids in the treatment of skin aging: an overview of clinical efficacy and safety. Clin Interview Aging. 2006;1(4):327-348
61. Kafi R, Kwak HS, Schumacher WE, Cho S, Hanft VN, Hamilton TA, King AL, Neal JD, Varani J, Fisher GJ, Voorhees JJ, Kang S. Improvement of naturally aged skin with vitamin A (retinol). Arch Dermatol. 2007;143(5):606-612
62. Rauma AL, Mykkanen H. Antioxidant status in

vegetarians versus omnivores. Nutrition 2000;16(2):111-119
63. Burke KE. Interaction of vitamins C and E as better cosmeceuticals. Dermatol Ther. 2007;20(5):314-321
64. Gupta M, Mahajan VK, Mehta KS, Chauhan PS. Zinc in dermatology: a review. Dermatol Res Pract. 2014;2014:709152
65. Hoffman L, Subramanyan K, Johnson AW, Tharp MD. Benefits of an emollient body wash for patients with chronic winter dry skin. Dermatologic Ther. 2008;21(5):416-421
66. Walters RM, Mao G, Gunn ET, Hornby S. Cleansing formulations that respect skin barrier integrity. Dermatol Res Pract. 2012;2012:495917 doi:10.1155/2012/495917 EPub 2012
67. Draelos Z, Hornby S, Walters RM, Appa Y. Hydrophobically modified polymers can minimize skin irritation potential caused by surfactant-based cleansers. J Cosmet Dermatol. 2013;12(4):314-321
68. European Patent Office. European patent application: application number 11194698.4, applicant Johnson & Johnson Consumer Companies Inc.; skin cleansing compositions with polyglycerol esters and hydrophobically modified polymers [Internet]. 2012 June [cited 2019 June]. Available from: https://patentimages.storage.googleapis.com/c2/c5/2c/e8733a7d2966e8/EP2468246A1.pdf

3. A Rosy Glow:

Red skin in the Cold

3. A Rosy Glow:
Red Skin in the Cold

After autumn turns the trees from green to gold, starts the smell of wood fires and beckons bobble hats out to bob along paths and pavements, winter brings cosy evenings, soft blankets and sparkly lights. Sudden changes in the temperature around us, hot drinks and heaters can also link pink cheeks and red noses with the start of cold weather. For some people this is just a subtle, momentary rosiness but for others it persists and crosses the line from a winter glow into an embarrassing problem.

Tiny blood vessels in our skin normally widen (vasodilate) when the body is too hot, allowing more blood to flow to the surface and body heat to be lost to our surroundings. This increase in blood flow causes our skin to flush pink or red. This is normal when we are wearing too many jumpers or drinking hot drinks, but skin can also flush for other reasons. In relation to cold weather, this chapter discusses a protective phenomenon called cold induced vasodilatation and a chronic skin disease called rosacea.

❖
Changes in temperature, hot drinks and heaters can link pink cheeks and red noses with the start of cold weather
❖

Cold Induced Vasodilatation (CIVD)

CIVD is a protective response designed to prevent frostbite. The temperature of our skin drops when it is exposed to cold weather and more so when it is exposed to cold, windy weather[1,2]. Our fingers, nose, cheeks and chin are often the coldest parts of our body and flow of blood to these areas changes in response to the temperature of the skin, producing obvious changes in skin colour[1].

We expect to see our skin flush when we are hot, but it is interesting to see it also turning red when we are out in the cold. If the redness means that we are losing body heat through our skin, when we are cold why does are body not divert blood away from the skin's surface, to conserve heat, with the reduction in blood flow causing the skin to become pale?

❖
CIVD is a protective response designed to prevent frostbite
❖

So, when we are cold our body does limit its loss of heat by restricting blood flow to the skin. Initially when we venture out into cold weather we might see our hands or feet lose a bit of their pink colour. This is caused by small blood vessels at the skin tightening (vasoconstriction). This only happens within the first few minutes of exposure to cold, and then it happens again later, if our body temperature was to drop below 37.5°C. At that later point, the body is so cold that it needs to sacrifice the flow of blood (which carries oxygen and nutrients) to the less important peripheral tissues in order to save our vital organs[1].

During the average uncomfortable exposure to cold weather, your body chooses to save your fingers and facial sticky-outy bits from the cell damage that occurs from restricted blood flow by using the process of CIVD.

The Physiology of CIVD

CIVD begins 5-10 minutes after the skin temperature at the peripheries drops below 15°C and usually by the time it reaches around 10°C[1,2]. The flow of blood to our skin is controlled by joins between small arteries and veins (called arteriovenous anastomoses), which respond to the drop in skin temperature by widening, opening up, increasing the flow of blood into the skin and increasing the skin temperature[2].

How much CIVD can increase the skin temperature by depends on how warm the body's core temperature is[1,2]. Taking part in moderate exercise in the cold keeps the fingers and nose warmer than they are when standing still, probably by the effect on the body's core temperature[1]. As a result of CIVD, the bits of the face that stick out, like the nose, stay warmer than other parts like the forehead or cheek[1,3]. But, if the core body temperature drops, CIVD stops, putting these peripheral areas at risk of cold injury[1,2].

❖

If the core body temperature drops, CIVD stops, putting peripheral areas at risk of cold injury[1,2]

❖

CIVD seems to be at least partly an adaptive response; a response that appears quickly and effectively if it is frequently needed, but more slowly and less effectively if it is not used that often. People who are repeatedly exposed to the cold, or living and working in very cold surroundings develop a faster, better CIVD response than those of us languishing in warmer environments[2,4]. Isn't that clever.

Rosacea

Complexions may be described as 'ruddy' - a term that may be used to describe pink or red skin in people who have spent years working outside. It may also be used when describing people who (with a dermatologists eye) have significant rosacea[5].

Rosacea is quite a common, chronic, facial skin condition within fair skinned populations. It can be mild, with occasional flare-ups noticed mainly by the individual, severe, with socially obvious and persistent signs, or somewhere in between.

❖
Rosacea is quite a common chronic facial skin condition
❖

Rosacea is actually an umbrella term for a wide range of facial skin changes including flushing, redness, dilated blood vessels (called telangiectasia), spots and pustules, itching and stinging, swelling and enlargement of the sebaceous tissue of the nose. Some features of rosacea can overlap with those of acne, perioral dermatitis and sun damage.

The variety of signs and symptoms in rosacea lead to the development of a classification system at the start of the century to avoid diagnostic confusion and make research into treatments easier to translate and apply. Rosacea was divided into four subtypes: erythemotelangiectatic, papulopustular, ocular and phymatous, and the diagnosis of rosacea is based on the following criteria[6]:

The presence of one or more of these main features:

Facial flushing, persisting redness, spots and pustules, telangiectasia.

And may include one or more of:

Burning, stinging, dry skin, swelling of the skin, dry or gritty eye symptoms, nasal skin changes.

❖
Rosacea is very visible and can cause significant embarrassment
❖

Not everyone will get all of the features. The commonest subtypes are erythemotelangiectatic and papulopustular rosacea, with phymatous rosacea being quite rare in comparison. In erythemotelangiectatic rosacea there is persisting redness in the centre of the face, visible blood vessels in the skin, frequent flushing and roughness in skin texture[6,7]. In papulopustular rosacea, people experience similar redness but also papules, pustules and inflamed nodules affecting the centre of the face. Papulopustular rosacea can overlap slightly with acne, but in rosacea the redness of the skin at the centre of the face is more pronounced and it lacks the comedones (blackheads) and scarring seen in acne[6,7].

Being a facial skin condition, rosacea is very visible and can cause significant embarrassment. Some people may manage to hide mild to moderate redness under green tinted creams or camouflage make up, but others are less able to do this. Severe rosacea, particularly inflamed papulopustular rosacea, is very difficult to hide and, like acne, covering it in a thick layer of make-up can make it worse.

Rosacea can have a tremendous impact on happiness and wellbeing, with research studies showing that it can affect people's emotional state, how they view themselves and how they live their lives[8]. People with severe rosacea may worry about how others will react to or interpret the condition in case it is assumed that the person is an alcoholic, lacking in confidence, or unhealthy in some other way[8].

What Causes Rosacea?

We don't know exactly why rosacea appears in some people and not in others. Genetics, disruption of the skin barrier, an oversensitivity in the skin's immune system, changes in the responses of blood vessels in the skin, hormonal changes, stress, imbalance of microbes and ultraviolet (UV) light may all play a role[7,9,10,11]. These and certain triggers for flare-ups can stimulate the release of inflammatory signalling molecules like cathelicidin, cytokines and chemokines from nerves, leading to vessels dilating and an increase in blood flow, flushing and skin sensitivity[9,11]. Frequent flushing can lead to permanently dilated, leaky blood vessels and persisting inflammation[12].

Triggers

> ❖
> Triggers for flushing and stinging contribute to the development and persistence of rosacea
> ❖

Triggers for flushing and stinging definitely contribute to the development and persistence of rosacea. Skin affected by rosacea is more sensitive to things that would not tend to cause a reaction in unaffected skin[11]. For example, people with rosacea flush more easily in response to heat compared to those without the condition[13]. Heat is a direct trigger for flushing, as even in unaffected skin it causes widening of blood vessels, with increased blood flow and redness. In rosacea, an exaggerated response to heat is probably caused by an oversensitivity in nerve endings and an overreaction in the signals they produce[13]. Moving from the cold into a warm room, sitting in a car with a heater directed at your face, emotional stress, hormonal changes, hot drinks, spicy food, alcohol, exercise, cold or hot weather, sunlight and hot baths or showers all commonly trigger the release of signalling molecules from nerves, leading to flushing and associated symptoms[9,10].

Sunlight is one of the most commonly described triggers for flushing and worsening of erythemotelangiectatic and papulopustular rosacea[10,11]. The UV rays in sunlight are felt to trigger flare ups by three processes: production of Vitamin D in the skin causing keratinocytes to over-produce cathelicidin (an important molecule in the defence against bacteria) leading to inflammation; UVB rays increasing the growth of new blood vessels in the skin; and the reactive oxygen species (ROS) generated in the skin on exposure to UV light contributing to ongoing inflammation[11].

Skin Barrier Dysfunction

Skin affected by rosacea is sensitive and easily irritated. Higher levels of water loss across the epidermis in affected skin indicate that there is a loss of integrity in the skin barrier[14]. Contact sensitivity, in which the skin produces an allergic or irritant response to a product or ingredient applied to it, is more common in people with rosacea compared to those without[15].

Disruption of the skin barrier makes it easier for potential contact allergens to cross, annoy the immune system and produce a reaction[14,15]. In susceptible people, contact sensitivity can develop from the use of cosmetics, washes and facial creams, 'natural' products and some ingredients of medical creams. Maintaining the integrity of the skin barrier with bland, well-designed skincare will help to prevent contact sensitivity from developing.

Microbial Imbalance

We all have bacteria and mites living on our skin but as long as their numbers stay within normal levels our skin doesn't tend to be bothered by them. Demodex folliculorum is a mite that is normally found on the skin, living essentially off of our dead skin cells, but in rosacea this mite is often seen in higher numbers within hair follicles[12]. The bacteria associated with these mites may contribute to immune activity, with release of inflammatory markers including cathelicidin driving inflammation and leading to changes in blood vessels[10,11,16].

Reactive Oxygen Species (ROS)

ROS are a normal by-product of energy generation, exercise and fighting infection and illness. While ROS are constantly being produced, they can damage cells and structures if not mopped up by the body's system of antioxidants. ROS are thought to be involved in the development of rosacea as higher levels of ROS are found in affected skin, and many medical treatments that are used successfully to treat rosacea reduce the levels of ROS[11].

Diet

Dietary triggers are common and have been grouped into four categories: heat related, alcohol related, capsaicin related (found in spices and peppers) and cinnamaldehyde related (found in tomatoes, citrus, cinnamon and chocolate)[10]. Culprit molecules may stimulate specific ion channels which then trigger and maintain symptoms like flushing, inflammation and sensitivity[10,17].

Diet may also impact rosacea through an as-yet undetermined relationship between the gut and the skin[10,18]. Our gut lining is a continuous layer of cells covering the inside of our mouth, throat, oesophagus, stomach, small and large intestines, rectum and anus. While our gut is generally considered to be 'inside' our bodies it is really another external surface, like the skin, a border between the outside world and our inner cells and organs.

Of the food that we chew and swallow, some is digested and absorbed across the gut lining for energy and nutrition, but significant amounts never see the true inside of our body as they pass through the sections of the gut and out in our poo. Like skin, the gut is susceptible to irritation and inflammation from some of the substances it comes into contact with, producing symptoms of irritation, intolerance or allergy. The gut can also develop inflammatory diseases like Crohn's Disease and ulcerative colitis.

There is an association between inflammatory bowel diseases like Crohn's Disease and ulcerative colitis and inflammatory skin diseases like rosacea, psoriasis and atopic dermatitis (eczema)[10,19,20]. People with rosacea may have a higher chance of having or developing inflammatory bowel disease or other gut diseases including coeliac disease and irritable bowel syndrome[18,19]. It is not clear at present why this is. It may due to similar cellular genetics at the two surfaces, the overall genetics of the individual, or related to the treatments of either disease increasing the chance of the other developing. It has been suggested that rosacea in a person with tummy troubles may be a sign of unrecognised gastrointestinal disease, and attention should be paid to any gastrointestinal symptoms in a patient with rosacea[10,18].

Bacterial overgrowth in the gut and infection with a bacteria called H. Pylori are more common in people with rosacea, with some reports of improvement in rosacea after successful treatment, but there is no firm scientific evidence of a link here[10,18]. These associations may be related to shared autoimmune characteristics or the frequency of antibiotic use in the management of rosacea, and research in the area continues.

If you take prolonged courses of antibiotics for rosacea, or even if you don't, it makes sense to try and protect your healthy gut microbiome by eating a wide range of plant-based foods, fibre, probiotics like live yoghurt and avoiding highly processed foods[10]. Ask your doctor for their opinion if you feel you suffer from rosacea and regular gastrointestinal symptoms like heartburn, indigestion, bloating, diarrhoea or constipation.

Managing Rosacea

At present there is no cure for rosacea. Management controls symptoms and tries to reduce the number of flare-ups, and is a combination of avoiding obvious triggers where possible, appropriate use of basic daily skincare and the use of prescription creams, tablets or other therapies when needed.

Trigger Avoidance

Triggers for flares of rosacea are often encountered in everyday life and can be difficult to avoid, but it is important to identify early on which triggers affect you, which ones you are particularly exposed to and how exposure can be reduced to a minimum. Car heaters can be angled toward the body rather than at the face and hairdryers, hot showers and hot drinks all reduced slightly in temperature.

> ❖
> Identify which triggers affect you and how exposure can be reduced to a minimum
> ❖

Exposure to sudden changes in environmental temperature is very difficult to avoid, particularly during cold weather. For many common triggers a balance may need to be met between the desire to avoid flushing and the desire to indulge in, for example, alcohol or spicy foods. People who spend more time outdoors are at higher risk of being affected by rosacea because of their frequent exposure to temperature fluctuations and UV light, but limiting UV through daily use of facial sunscreen, and limiting the contact between direct heat and facial skin as above may help.

Facial Skincare

While some people may need specialist medical treatment, everyone who suffers from rosacea will benefit from starting with appropriate daily skincare. Unsuitable, harsh basic skincare can trigger or worsen signs and symptoms of erythemotelangiectatic and papulopustular rosacea, but mild cleansers and moisturisers will work with any prescribed treatment by encouraging skin barrier repair[7,9].

The facial skincare market is huge. There is an overwhelming volume of products on offer, all promising effects that they may or may not be able to deliver. Because skin affected by rosacea is sensitive it can be difficult to know which face washes and moisturisers to go for.

❖
Everyone who suffers from rosacea will benefit from starting with appropriate daily skincare
❖

To protect the skin barrier and avoid irritation, many clinicians advise people with rosacea to avoid using toners, abrasive exfoliators or dermabrasion, acidic peels, thick, heavy cosmetics or sensory stimulants such as menthol or camphor[7,12]. Avoidance, or care when using products containing acetone, alcohol, propylene glycol, alpha hydroxy acids (AHAs), sodium laureth sulphate (SLS), benzylkonium chloride, formaldehyde releasers (including quaternium-15, DMDM hydantoin), benzoyl alcohol, urea, pyrrolidone carboxylic acid (PCA), lanolin and fragrances has also been advised, though these may be tolerated if present in small amounts[7,12].

As discussed in Chapter 1, many anionic, harsh cleansers can damage the stratum corneum by stripping it of lipids, proteins and natural moisturising factor, making the skin more sensitive. Cationic combars may make dryness or irritation worse and are not suited for patients with rosacea[7]. Alkaline cleansers like soap can upset the pH balance at the skin surface with consequences for the healthy microbiome and skin barrier[21,22,23,24]. Syndet and soap-free cleansers are best for use in rosacea, but in these products the surfactants (the ingredients responsible for removing oil and debris) used will determine whether or not a product will produce irritation in a given individual[7,21]. Nonionic and amphoteric syndet cleansers are less irritating and drying and are designed to clean with minimal stripping of essential lipids and proteins. Preparations may contain their own lipids like vegetable oils, petrolatum, ceramides or cholesterol to try and reduce any damage[7,21].

> ❖
> Gentle syndet and soap free cleansers are best for use in rosacea[7,21]
> ❖

The right ingredients in a moisturiser can help to repair the skin barrier. People with erythemotelangiectatic or papulopustular rosacea should use products that increase skin hydration and do not cause irritation. Depending on concentration, urea can cause stinging, propylene glycol may produce irritant dermatitis, alpha hydroxy acids (AHAs) like glycolic or lactic acid may produce stinging and irritation and lanolin can cause allergic or irritant reactions[7]. Try fragrance-free products designed for sensitive skin, considering the ingredients described in Chapter 2.

Suggested Daily Care[7,9,12]:

- Gently wash the face morning and night with lukewarm water and a syndet or soap-free cleanser intended for sensitive skin.
- Use an oil-free, broad-spectrum sunscreen everyday, SPF at least 15. Mineral sunscreens with titanium dioxide or zinc oxide are usually well tolerated. Further protection with a hat may be needed on especially sunny days.
- Use a therapeutic, non-comedogenic or non-acnegenic moisturiser after gently washing and drying the face. Anti-inflammatory ingredients like nicotinamide may be helpful.
- Use mineral or water based, non-comedogenic, non-acnegenic make up.

Medical Treatments

If rosacea persists despite appropriate daily skincare there are treatments that your GP or dermatologist can prescribe to try to improve things. These can be topical (applied to the skin) or oral (taken by mouth). They can be used alone or in combination and can produce good effects, but unfortunately some can cause skin irritation. If the irritation is significant then treatment should be changed, but if it is mild then it may be relieved by adjustments to your daily skincare regime, or in some cases by trying it every second or third day to begin with and increasing its use when possible.

Topical Treatments

The spots, pustules and redness from inflammation in papulopustular rosacea are often the best responders to topical treatment, though some redness in erythemotelangiectatic rosacea can also respond[9]. Options are described in the table below (these may vary depending on local guidelines):

Topical Treatment (UK, 2019)	Action[25]
Metronidazole cream / gel	Reduces inflammation including redness, papules and pustules.
Azelaic acid cream / lotion	Reduces inflammation, redness and lesions.
Tacrolimus or pimecrolimus ointment	Reduces inflammation, redness and lesions.
Ivermectin cream	Lowers the volume of demodex mites, may reduce inflammation and lesions.
Retinoid cream	Reduces redness and lesions.
Brimonidine gel	Reduces redness and flushing by a temporary tightening of blood vessels.

Avoid applying topical steroids to skin affected by rosacea. In the short term it can improve redness and irritation, but in the longer term it makes things worse.

Oral Treatments

Oral treatments may be tried in people who have not noticed an improvement in basic and topical treatment, and can be used in combination with some topical treatments, like azalaic acid. Again, papulopustular rosacea is the most responsive to most of these treatments, and options may vary depending on local guidelines.

Oral Treatment (UK, 2019)	**Action**[25]
Oral antibiotic such as a tetracycline or erythromycin for 3-6 months	Reduces inflammation, redness and lesions.
Isotretinoin (dosed by body weight)	A retinoid drug. Reduces inflammation, adjusts the activity of the skin's immune system. Isotretinoin can sometimes reduce the swelling and volume of rhinophyma[26]

Oral antibiotics have been used to treat rosacea for several decades, usually tetracyclines but occasionally macrolides, and are believed to produce their effect by reducing inflammation. Doxycycline is commonly used, either at its full antibiotic dose or at a sub-antimicrobial dose[7,9]. Treatment is continued for 3-6 months but this can produce side effects as well as having the potential for increasing bacterial resistance. Intermittent courses of antibiotic are often required to control flares. Using a prescribed non-antibiotic topical treatment alongside, and continuing this in the longterm after oral treatment has stopped, can produce a better response and help to maintain it for longer.

Isotretinoin (an oral retinoid drug) may be needed for severe papulopustular rosacea to reduce inflammation and inflammatory lesions. It can also shrink the early changes of phymatous rosacea, probably by reducing the size of sebaceous glands[26]. Isotretinoin is usually taken for several months, but rosacea can return after this course and so in some cases the drug is used at a low dose for months to years[9,26]. As a retinoid, isotretinoin has important potential side effects and is not suitable for everyone, particularly for longterm treatment.

Laser or Light Based Treatments

The persisting dilated blood vessels (telangiectasia) seen in erythemotelangiectatic rosacea can only really be successfully improved with laser, light or other coagulant treatments. Intense pulsed light (IPL) can reduce the redness surrounding lesions, while pulsed dye (PDL) or Nd:YAG lasers may be used to treat telangiectasia[25]. Lasers release light energy that is absorbed by the haemoglobin inside blood vessels, leading to heating of the vessel and it closing up. Laser may improve inflammatory lesions, though it can also make inflammatory lesions worse[25]. One process of management would be to treat inflammatory papules and pustules first with medical treatment, then move on to treating redness and telangiectasia with laser afterwards. CO_2 laser or surgery may be used to treat rhinophyma.

Depending on local protocols, apart from in very severe cases these treatments may be considered cosmetic and take place in the private sector. Always go to a reputable, suitably trained provider.

Summary

- Red skin in cold weather is common and may be a normal part of our physiology or may be related to skin disease.
- Cold induced vasodilatation is designed to prevent frostbite and produces redness at our fingers, noses and other exposed areas while out in the cold.
- Rosacea is a common, chronic, visible skin condition that can cause significant embarrassment to people affected by it.
- Rosacea can be managed by measures including using appropriate daily skincare, avoiding triggers as far as possible and protecting facial skin from the sun.
- If this fails, GPs or dermatologists can discuss and prescribe suitable treatment.

References

1. Gavhed D. Human responses to cold and wind. PhD Thesis. Department of Physiology and Pharmacology, Karolinska Institutet, Stockholm, Sweden. 2003
2 Daanen HA. Finger cold-induced vasodilatation: a review. Eur J Appl Physiol. 2003;89(5):411-426
3 Brajkovic D, Ducharme MB. Facial cold-induced vasodilatation and skin temperature during exposure to cold wind. Eur J Appl Physiol. 2006;96(6):711-721
4 Lee JY, Park J, Koh E, Cha S. Finger cold-induced vasodilation of older Korean female divers, haenyeo: effects of chronic cold exposure and ageing. Int J Biometeorol. 2017;61(7): 1299-1307
5 Vocabulary.com. 'Ruddy' [internet]. United States of America: Vocabulary.com; [updated unknown; cited 2019 June]. Available from: https://www.vocabulary.com/dictionary/ruddy
6 Wilkin J, Dahl M, Delmar M, Drake L, Feinstein A, Odom R, Powell F. Standard classification of rosacea: report of the national rosacea society expert committee on the classification and staging of rosacea. J Am Acad Dermatol. 2002;46(4):584-587
7 Levin J, Miller R. A guide to the ingredients and potential benefits of over-the-counter cleansers and moisturizers for rosacea patients. J Clin Aesthet Dermatol. 2011;4(8):31-49
8 Zeichner JA, Eichenfield LF, Feldman SR, Kasteler JS, Ferrusi IL. Quality of life in individuals with erythemotelangiectatic and papulopustular rosacea: findings from a web-based survey. J Clin Aesthet Dermatol. 2018;11(2):47-52
9 Rainer BM, Kant S, Chien AL. Rosacea: epidemiology, pathogenesis and treatment. Dermatoendocrinol. 2018;9(1):e1361574
10 Weiss E, Katta R. Diet and rosacea: the role of dietary change in the management of rosacea. Dermatol Pract Concept. 2017;7(4):31-37
11 Yamasaki K, Gallo RL. The molecular pathology of rosacea. J Dermatol Sci. 2009;55(2):77-81
12 Bolognia JL, Jorizzo JL, Schaffer JV. Dermatology. 3rd ed. United States of America: Elsevier Saunders; 2012
13 Guzman-Sanchez DA, Ishiuji Y, Patel T, Fountain J, Chan YH, Yosipovitch G. Enhanced skin blood flow and

sensitivity to noxious heat stimuli in papulopustular rosacea. J Am Acad Dermatol. 2007;57(5):800-805
14 Dirshka T, Tronnier H, Forster-Holst R. Epithelial barrier function and atopic diathesis in rosacea and perioral dermatitis. Br J Dermatol. 2004;150(6):1136-41
15 Erdogan HK, Bulur I, Saracoglu ZN, Bilgin M. The evaluation of contact sensitivity with standard and cosmetic patch test series in rosacea patients. Ann Dermatol. 2018;30(3):290-295
16 Baumann L, Saghari S, Weisberg E. Cosmetic dermatology: principles and practice. 2nd ed. United States of America: McGraw-Hill Companies Inc.; 2009
17 Sulk M, Seeliger S, Aubert J, Schwab VD, Cevikbas F, Rivier M, Nowak P, Voegel JJ, Buddenkotte J, Steinhoff M. Distribution and expression of non-neuronal transient receptor potential (TRPV) ion channels in rosacea. J Invest Dermatol. 2012;132(4):1253-1262
18 Egeberg A, Weinstock LB, Thyssen EP, Gislason GH, Thyssen JP. Rosacea and gastrointestinal disorders, a population based cohort study. Br J Dermatol. 2017;176(1):100-106
19 We CY, Chang YT, Juan CK, Shieh JJ, Lin YP, Liu HN, Lin JT, Chen YJ. Risk of inflammatory bowel disease in patients with rosacea: results from a nationwide cohort study in Taiwan. J Am Acad Dermatol. 2017;76(5):911-917
20 Kim M, Choi KH, Hwang SW, Lee YB, Park HJ, Bae JM. Inflammatory bowel disease is associated with an increased risk of inflammatory skin diseases: a population-based cross-sectional study. J Am Acad Dermatol. 2017;76(1):40-48
21. Draelos ZD. The science behind skin care: cleansers. J Cosmet Dermatol. 2018;17(1):8-14
22. Ali SM, Yosipovitch G. Skin pH: from basic science to basic skin care. Acta Derm Venereol. 2013;93(3):261-267
23. Proksch E. pH in nature, humans and skin. J Dermatol. 2018;45(9):1044-1052
24. Draelos Z, Hornby S, Walters RM, Appa Y. Hydrophobically modified polymers can minimize skin irritation potential caused by surfactant-based cleansers. J Cosmet Dermatol. 2013;12(4):314-321
25. Abokwidir M, Feldman SR. Rosacea Management. Skin Appendage Disord. 2016;2(1-2):26-34
26. Park H, Del Rosso JQ. Use of oral isotretinoin in the

management of rosacea. J Clin Aesthet Dermatol. 2011;4(9):54-61

4. Stocking the Hibernaculum: Food and Skincare

4. Stocking the Hibernaculum:
Food and Skincare

Whether it's because of a physical need for insulation, a comfort encouraged by dark evenings or that we feel happier indulging under layers of clothing, cold days feel better with good things to eat and drink.

There is centuries of artistic and scientific evidence to support the impact of our diet on our health and the way we look. A balanced diet with the necessary calories will help to keep us healthy, but there is also a growing market and some scientific evidence for some specific elements in our food being particularly important for the health of our skin. Some of these may help to improve common cold weather skin problems, giving us an extra excuse to eat them.

But, no one magic ingredient will cure skin problems, and overdosing on any one item can be as dangerous as deficiency. Most of the elements discussed in this chapter are best included as part of a healthy, balanced diet.

Choosing Food

At present in the UK we live in luxury when it comes to food choice. Many foods that would be seasonal are available all year round and ingredients from all over the world, covering many cultures and diets, can be bought easily from the local corner shop or supermarket. More of us travel internationally, eat out and scroll through social media and we are aware of far more food options than our grandparents would have considered.

Expansion in food choice has improved the lives of people living with medical conditions like coeliac disease, food allergies and diabetes, who can enjoy a tasty and varied diet while remaining symptom free. Beyond medical need, variety means that we can be picky with our tastes and tailor our diets in a way that was not possible 50-100 years ago.

❖

Variety means that we can be picky with our tastes and tailor our diets

❖

In the 1940's, people ate seasonally and bought food from local butchers, bakers and grocers. They grew more of their own food and even urban households grew up to half of their own fruit and vegetables in the summer[1]. Things changed gradually over the following decades because of relief from rationing, higher numbers of women in fulltime work, developments and affordability of kitchen equipment and the emergence of supermarkets[1]. The drive towards variety and convenience foods at affordable prices has obvious advantages but can impact on our environment, the farming industry, public awareness of food production processes and possibly our health.

The UK has imported and exported food for centuries[2]. In the last few years, most of the food eaten in the UK was produced here but supermarkets and suppliers also ensured our food security and catered to our demands by importing almost a third of food from the European Union and about a fifth from the rest of the world[3,4]. The monetary value of the fruit and vegetables that we import has been the highest out of all imported food groups[4]. Apples, pears, strawberries, blackberries, blueberries, plums and many other fruits can be grown well in the UK but tend to be limited by their seasonal growth and have been felt to be labour intensive to harvest[2].

Choosing sustainable, quality food when we shop encourages this to be produced

Food Quality

Supermarket competition and low cost retailers can put farmers under pressure on trading terms and the price they receive for their produce. As a result it may be difficult for farmers that choose to invest in higher environmental, labour or animal welfare standards to balance their books and enjoy a sustainable livelihood[2]. Choosing sustainable, quality food when we shop encourages this to be produced and supports those farmers who are working hard to provide quality, ethical produce.

However, there is often a significant price difference between healthy and unhealthy food, with healthy foods being more expensive[5]. When there is little money to spend, households may buy unhealthy foods that are filling and will not go to waste, increasing their risk of obesity and related health problems[6]. It is important for a balance to be achieved to ensure that we all have the ability to eat healthy, nutritious food, while maintaining quality standards and a livelihood for our hardworking farmers.

Sugar

During cold, wintry weather our diet is often rich in sugar. Chocolate truffles, chocolate oranges, chocolate advent calendars, hot chocolate, candy canes, mince pies, champagne, prosecco… there is even considerable amounts of sugar where you wouldn't really expect it, like in stuffing and mulled wine[7]. It can be difficult to get away from and often too tasty to refuse. But, diets high in sugar can produce high sugar levels in the blood and in the skin, leading to changes in skin structure and appearance[8].

The NHS website advises limiting intake of free sugars to not more than 30g per day for those over the age of 11, maximum 24g for those aged 7-10, or 19g for those aged 4-6[9,10]. Free sugars are the sugars added to foods and drinks including cakes, biscuits, chocolate and fizzy drinks, and those found naturally in honey, syrups, fruit and vegetable juices and smoothies. Sugars found naturally in fruit, vegetables and milk are not free sugars and we do not need to reduce our intake of these[10].

Glycation and Advanced Glycation End Products (AGEs)

A body process called glycation has been widely studied in the context of ageing[11]. Sugar molecules from our diet amble around the body until they find suitably attractive proteins, lipids or nucleic acids to bind to. The process of binding is called glycation. Glycation produces advanced glycation end products (AGEs) and high numbers of AGEs gathering in the body have been linked to many diseases[11]. In the skin, glycation results in specific structural and functional changes that have been termed 'sugar sag'[11]. Glycation and AGE production speeds up in the presence of high sugar levels, reactive oxygen species (ROS) and ultraviolet (UV) light[8,11].

❖
Glycation results in specific changes that have been termed 'sugar sag'[11]
❖

Collagen, which is so important for skin's structure, is affected by glycation in several ways. It becomes stiff and vulnerable to mechanical stress, less able to interact with surrounding cells and proteins, less able to maintain its structure, and becomes more resistant to degradation, which stops it from being replenished with new proteins[11]. High levels of AGEs also affect elastin, which is needed for normal cell functions and for skin to stay stretchy[11].

Limiting AGEs

As well as monitoring our free sugar intake, altering our cooking methods can help reduce the production of AGEs in our bodies. Methods associated with producing cold weather comfort food like frying and roasting are associated with high AGE formation, whereas boiling and steaming produce far less[8].

Some molecules and elements in our diet may limit AGE production and the damage they may cause, including flavonoids, nicotinamide and zinc[11]. Glycation speeds up in the presence of ROS, so antioxidants should in theory help to reverse this[11]. Resveratrol (found in red wine and in the skin of grapes, blueberries, raspberries and mulberries) and L-carnitine (found in fish, dairy, meat and poultry) are both antioxidant compounds that may limit AGE formation and related tissue damage[11].

Interestingly, several spices that tend to be associated with the cold weather season are thought to be able to limit production of AGEs, including cinnamon, cloves and allspice[11].

Fat

Typical western pattern diets are rich in protein and fats, while vegetarian and vegan diets are generally lower in fat and cholesterol[12]. While too much fat can be bad for us some fat is essential for daily life. We need essential fatty acids for many body functions and fat helps with the absorption of vitamins A, D and E across the gut. NHS Choices recommends men should limit their saturated fat intake to 30g per day, women to 20g per day, and children should 'have less'[13].

Saturated and Unsaturated Fat

Fats are made from chains of fatty acids held together with chemical bonds and may be saturated or unsaturated depending on their chemical structure. High levels of saturated fat are found in dairy items like cream, cheese and butter, and also coconut oil and processed meats[13]. Saturated fatty acids include lauric acid, palmitic acid, stearic acid and butyric acid.

Replacing saturated fats with unsaturated ones may lower levels of 'bad' cholesterol[13]. Saturated fats tend to be included in processed foods, partly because unsaturated fats are more vulnerable to rancidity, or going-off, than saturated ones are[14]. Unsaturated fats can be monounsaturated, such as those found in olive oil, avocado and nuts; or polyunsaturated, which include omega-3 fatty acids found in oily fish, and omega-6 fatty acids found in rapeseed and sunflower oils[13]. Examples of unsaturated fatty acids include oleic acid, linoleic acid, arachidonic acid and palmitoleic acid.

Polyunsaturated Fatty Acids (PUFAs)

PUFAs include omega-3 and omega-6 fatty acids[13]. They are involved in signalling pathways managing a wide range of bodily processes, including inflammation[12,15,16,17]. They help to keep skin hydrated by contributing to the structure of the lipid bilayer in the stratum corneum.

Fish oils containing omega-3 and -6 fatty acids are felt to reduce inflammation, improve symptoms of skin dryness and irritation, encourage wound healing and help to prevent sunburn and sundamage[17,18]. Omega-3 acids include alpha linolenic acid (ALA), eicosapentaenoic acid (EPA) and docosahexaenoic acid (DHA), while omega-6 acids include linoleic acid, gamma-linolenic acid (GLA) and arachidonic acid. The most common fatty acids in the skin are the omega-6 polyunsaturated fatty acids, linoleic and arachidonic acid[19].

❖
PUFAs help to keep skin hydrated by contributing to the structure of the lipid bilayer
❖

The best sources of omega-3 EPA and DHA in the diet are fish, fish oils and seafood[12,13,17]. Oils from mackerel and sardine have significant amounts of EPA and DHA and omega-3 eggs can provide some DHA[16]. Flaxseed oil is rich in ALA, which can be converted within the body to create EPA and DHA, though not as much as when they are consumed directly[12]. Marine algae and seaweed are also sources of ALA, which may be converted into low levels of EPA and DHA as above[12,20].

Omega-6 fats are found in vegetable oils such as rapeseed and sunflower oil and some nuts[13]. Borage seed oil is rich in GLA[21]. It has been advised that western pattern diets generally contain enough omega-6 fatty acids without a need to eat any extra, and very high levels of omega-6 fatty acids have been linked to high levels of plasma prostaglandin E2, which can encourage inflammation and suppress the immune system[13,15].

Vegetarian and especially vegan diets are relatively low in ALA and provide little to no EPA and DHA[12]. There is evidence that tissue levels of omega-3 fatty acids are lower in vegetarians and especially vegans, which may lead to drier skin[12,22]. Vegetarians and vegans may benefit from increasing their intake of EPA and DHA where possible, but this may be difficult[12,22].

Cholesterol

> ❖
> With lower cholesterol, vegetarians may be more prone to developing dry skin[22]
> ❖

Cholesterol is an essential component of cell walls and is involved in cell signalling, nerve conduction and the generation of hormones including vitamin D. In skincare, cholesterol is important in maintaining balance between the oiliness and dryness of skin. In addition to the omega-3 issue above, with lower cholesterol, vegetarians may be more prone to developing dry skin[22].

Cholesterol is produced in the liver and transported around the body attached to molecules called lipoproteins, which can be high density (HDL), low density (LDL) or very low density (VLDL). HDL carries cholesterol away from cells and back to the liver, where it is broken down and passed out of the body as waste. HDL is referred to as 'good' cholesterol, and higher levels are better[23]. LDL carries cholesterol to cells that need it but if there's too much for the cells to use it can build up in the walls of arteries, leading to the inflammation and narrowing of arteries known as atherosclerosis.

Dietary sources of cholesterol include cheese, beef, pork, egg yolk, poultry and fish. Dietary cholesterol actually has little impact on blood cholesterol, it is your saturated fat intake that is more important to monitor in terms of general health[23]. High blood cholesterol levels can be caused by diets high in saturated fat, little exercise, high BMI, smoking (which stops HDL from transporting cholesterol to the liver), high alcohol intake, and certain genetic conditions.

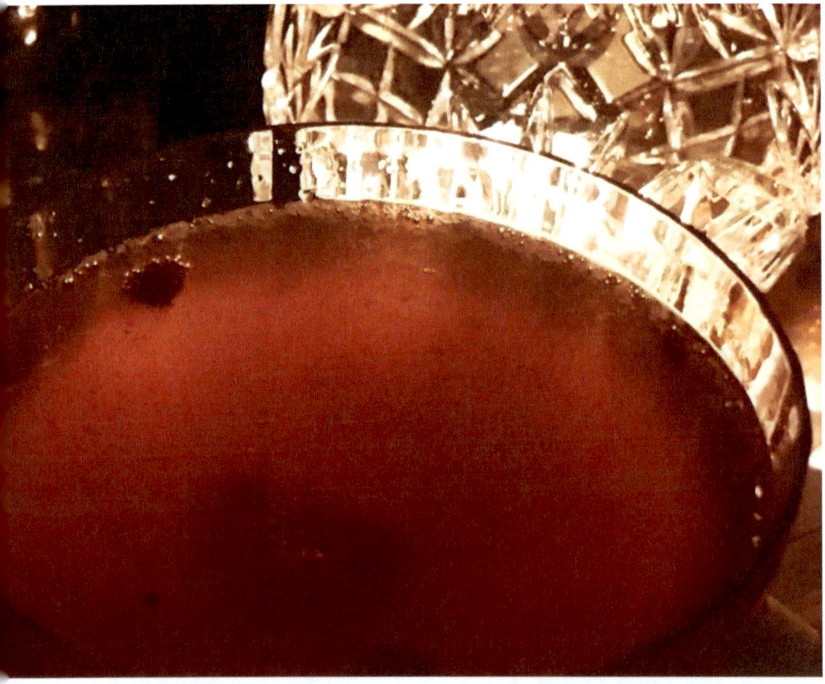

Antioxidants

As discussed in Chapter 1, antioxidants are a group of vitamins, pro-vitamins, enzymes, minerals, molecules and compounds, including vitamins A (retinoids and carotenoids), C and E (tocopherols), as well as zinc, copper, selenium, glutathione, co-enzyme Q10 and polyphenol compounds[24,25]. Antioxidants have a vital role in protecting our cells and DNA by mopping up potentially harmful by-products of essential body processes, like the ROS and free radicals generated by oxidative phosphorylation.

A state called oxidative stress occurs when by-products like ROS overwhelm the mopping-up ability of our antioxidant systems and the resulting damage to proteins, lipids and DNA can lead to cell death, replication of damaged genes and / or the development of cancer. There is almost no area of human disease in which oxidative stress has not been associated[26].

There is growing scientific evidence supporting the use of antioxidants at times when our own systems are overwhelmed by oxidative stress or declining through ageing or disease[25]. As has happened with vitamin A and synthetic retinoids, it is possible to create powerful medications from antioxidants. In skincare, antioxidants may be useful in reducing inflammation, in the prevention of sun damage and possibly in the prevention of skin cancer[25]. Vitamins C and E, polypodium leucomotos, lycopene, lutein, epigallocatechin gallate, resveratrol and co-enzyme

Q10 are examples of antioxidants with potential roles in skincare[25].

Antioxidants in Our Diet

Sources of antioxidants in our diet include berries, fruits, vegetables, beans, roots, cereals, nuts and seeds[27]. In the UK many people eat less fruit and vegetables than public health advice recommends, so may be lacking in important antioxidants and fibre[2,28]. Generally, vegetarians and vegans eat plenty of fruit and vegetables, cereals, pulses and nuts, and so their diet is well placed to provide enough antioxidant vitamins[24].

Many of us take vitamin and mineral supplements. Supplements are often felt to be free from significant side effects and to be 'natural substances', even though today they are mostly synthetic[29]. But some vitamins can cause unwanted effects. Taking high levels of vitamin A during pregnancy can cause significant problems with the baby's development (called teratogenicity), and ingesting very high amounts of vitamin A at any time can lead to liver and bone damage, acute or chronic poisoning, and even death[29].

The Oxidant Balance

There is a dynamic balance between oxidative phosphorylation, generation of ROS and free radicals, and antioxidants. ROS are generated during normal, essential processes like breathing, exercise and fighting infections and attract the attention of antioxidants, which act like a brake or buffer on their activity. Antioxidants act together as part of an interconnected system. Some work in synergy, like vitamins C and E, while others provide support in other ways, like vitamin E regenerating beta-carotene and protecting vitamin A from oxidation[25,29,30]. Over-supplementation with individual antioxidants can upset the balance of the system, potentially producing the opposite of the effect intended.

❖
Antioxidants act together as part of an interconnected system
❖

We can support our antioxidant systems by eating a healthy varied diet, including foods containing antioxidants of the kinds normally present in the skin. High doses of isolated antioxidants should be avoided, as should eating a lot of supplements in the place of normal food[31,32]. The aim should be for our body to have enough antioxidants, vitamins and minerals to fulfil its intended functions, not to swamp the system with one or two particular molecules. If taken in supplement form, consuming antioxidants in smaller doses and in combination has been described as the safest method of their use as this limits the risks of toxicity and potential drug interactions[25].

Nutriceuticals for Cold Weather

This section discusses specific elements of our diet that may help our skin. In relation to the effects of cold weather, many have been suggested to improve the hydration of our skin, reduce inflammation or protect our skin from the effects of intermittent strong sunlight. The inventory is not exhaustive, advice in this area may change as research progresses, and dietary manipulation should only be used to complement recommended medical treatment, not in place of it.

Where possible, while some elements may only be easily available as supplements, altering choices of foods to include these elements as part of a healthy diet is better than adding supplements on top of a poor diet[31,32]. If supplements are taken, they should always be obtained from reliable sources, taken as directed, and with attention to any potential side effects or drug interactions. Some can cause harm in high doses.

Nutriceutical Inventory

Biotin

Sources[33]:	Sweet potato, spinach, green beans, mushrooms, cheeses, cauliflower, nuts, eggs.

Biotin is also known as vitamin B7. Taking biotin daily for 6-15 months can increase nail thickness in people with brittle nails, but may not work as well in everyone[34]. It takes about 2 months to notice a difference, and unfortunately nail brittleness may return about 10 weeks after stopping treatment. Dermatologists may suggest biotin supplements to patients seeking advice for nail problems. It is not clear whether the supplementation treats a deficiency, or acts by another mechanism[34].

Carotenoids (see also Vitamin A, Retinol)

Sources[35,36,37]:	Yellow, orange, red and green coloured fruit and vegetables eg. mangoes, apricots, carrots, sweet potatoes, pumpkin, squash, melon, tomatoes, tomato products, broccoli, spinach, kale. Goji berries, chilli powder, paprika.

Carotenoids, including beta-carotene, lycopene and lutein, are pigments produced by some plants and algae. Once eaten they can be converted into biologically active retinoids, initially retinaldehyde (retinal) in the intestine and then retinol or all-trans retinoic acid in the body when needed[38,39].

Dietary supplementation with beta-carotene is felt to be safer than taking retinol and it may protect against sun damage and sunburn, but there are still safety concerns at high doses, when it acts as a pro-oxidant and can encourage DNA damage[29,40]. There is evidence that supplementation at high doses of 20-30mg/day can lead to an increase in lung cancer in smokers and similar results were seen in people with a history of exposure to asbestos[41,42]. Beta-carotene is often present in multivitamins, at safe doses of around 0.3mg, far lower than the high doses recorded in these trials[42]. The dose should always be listed on the packaging. Carotenemia (a yellow tinge to the skin) can develop when too much beta-carotene accumulates in fatty tissue.

Lycopene is the pigment mainly responsible for the red colour of tomatoes, and has potent antioxidant activity[40,43]. In combination with other carotenoids, such as when it is eaten as tomato paste, carrot juice, tomato extract, or in supplement form, lycopene seems to lower the risk of sunburn and higher levels of skin lycopene have been linked to smoother skin[40]. Lutein, another carotenoid, may also improve skin health, may reduce the risk of skin cancer and may reduce wrinkles[40].

Co-Enzyme Q10 (Ubiquinol)

Sources[44]:	Fish including sardines and mackerel, shellfish. Spinach, nuts. meat, some oils.

Co-enzyme Q10 assists with the production of energy in cell mitochondria. It is thought to prevent cell damage and cell death by limiting peroxidation of lipids, limiting free radical production and oxidative stress[45]. It is a powerful antioxidant in one form, known as ubiquinol, but can also exist in another form, ubiquinone, which is a pro-oxidant[26,45,46].

The antioxidant effect of ubiquinol is roughly the same as that of vitamin E (alpha tocopherol) and it also regenerates vitamin E[45]. Ubiquinol is the only lipid-soluble antioxidant that animal cells can make themselves and for which there is a mechanism for regeneration, though the activity of this reduces over time. Co-enzyme Q10 levels in cells reduce with advancing age and lack of exercise[45].

Glucosamine

Sources:	Supplements

Glucosamine is present in all human tissues. Glucosamine and its derivative N-acetyl glucosamine are particularly involved in the production of cartilage and hyaluronic acid, hyaluronic acid being an important molecule for skin hydration[47]. Much of the literature on glucosamine focuses on its use in the management of arthritis, but there are some studies examining the benefits on skin hydration and wrinkle reduction[47].

By increasing the production of hyaluronic acid, glucosamine may improve skin hydration, reduce wrinkles and fine lines and speed up wound healing[47]. A study published in 2002 found that 60 days treatment with N-acetyl glucosamine significantly increased skin hydration in women (ages teens to 30's) with usually dry, rough skin[48].

Hyaluronic acid (HA)

Sources:	Supplements

Hyaluronic acid can attract water up to 1000 times its own volume and plays important roles in skin hydration, cell growth, cell membrane function and cell adhesion. Oral supplements are available, but hyaluronic acid is metabolised (broken down) in the stomach and there is no evidence that enough is absorbed to be effective[22]. Hyaluronic acid is a common and effective ingredient of several injectable dermal fillers.

Iron

Sources[36,37]:	Red meat, spinach, dried beans, dried fruits, egg yolks, salmon, tuna, whole grains.

Iron is important for general health and is essential in the production of haemoglobin, which carries oxygen around the body in the blood. Low iron levels and iron stores have been associated with hair loss, dry and brittle hair, itchiness and dry skin, nail changes and cracking at the corners of the mouth. Some iron can be obtained through multivitamins, but if iron levels are significantly low (such as in anaemia) then your doctor may advise specific iron supplements.

Nicotinamide

Sources[49]:	Yeast, oat, maize, wheat, palm kernel oil, soya beans, molasses, liver

Nicotinamide is the amide form of niacin (vitamin B3) and also known as niacinamide. It is a synthetically produced water-soluble vitamin. It cannot be formed in the body so must be obtained through diet or supplementation. Nicotinamide is used to treat a variety of inflammatory skin conditions, including rosacea, acne, blistering skin conditions and sun damage[50]. Depending on its concentration, it may reduce itch, combat infection, protect against the reduction in the skin's immune system caused by sunlight, reduce sebum production and reduce uneven skin pigmentation[49]. It speeds up production of ceramides, which can improve skin barrier function and stabilises histamine containing mast cells, preventing them from bursting and releasing their contents[49]. It has few side effects and is safe, but there is little clinical data available on its effectiveness[49].

Omega-3 and -6 Fatty Acids

Sources[12,13,16,20]:

> Omega-3: ALA in fish oils like cod liver oil, marine algae, seaweed, walnuts, omega-3 eggs. EPA and DHA in fish like mackerel and sardine, other seafood and omega 3 eggs.
> Omega-6: GLA in evening primrose oil, rapeseed oil, borage seed oil, green leafy vegetables and nuts

---

Polyunsaturated fatty acids (PUFAs) are crucial components of cell membranes and the skin barrier. Alpha linolenic acid (ALA), eicosapentaenoic acid (EPA) and docosahexaenoic acid (DHA) are omega-3 fatty acids[15], while omega-6 acids include linoleic acid, gamma-linolenic acid (GLA) and arachidonic acid. EPA and DHA can be made in the body from intake of ALA, but not much is made in comparison to when they are eaten directly, so it is best to include them in the diet when possible[12].

The anti-inflammatory activity of omega-3 fatty acids is well established, and several studies have demonstrated their ability to improve the redness and irritation caused by psoriasis, rosacea, and sunburn[15,16,51]. The hydrating qualities of omega-3 fatty acids may help to add volume to the skin and reduce dryness, itch and the appearance of fine lines[17,18,52]. Omega-3 PUFAs may also protect the skin from cell and DNA damage caused by sunlight, by reducing inflammation and immunosuppression in exposed skin[15,17].

But, being unsaturated fatty acids, omega-3 PUFAs are unstable and liable to undergo lipid peroxidation. Their susceptibility to attack by free radicals may protect other structures nearby, but in order to protect our tissues from the effects of free radicals and lipid peroxidation it has been advised that decent amounts of antioxidants should be eaten alongside sources of omega-3[53].

Regarding omega-6 fatty acids, one study in elderly patients found that a daily dose of borage seed oil (rich in GLA) for 2 months significantly improved skin barrier function and reduced dry skin complaints[21]. Evening primrose oil contains both linoleic acid and GLA, and may improve skin hydration, reduce inflammation, and improve symptoms of atopic eczema, including inflammation, dryness and itch[54]. It has been used in combination with zinc to soothe dry eyes and improve brittle nails[22].

However, high omega-6 fatty acid intake has been linked to high levels of a molecule called prostaglandin E2, which encourages inflammation and immunosuppression and may potentially increase the chance of cancerous changes from exposure to sunlight[15]. Larger trial data suggests that lowering intake of omega-6 fatty acids may reduce the risk of basal and squamous cell skin cancer, while increasing intake of omega-3 fatty acids may give protection from sunburn and reduce the risk of squamous cell skin cancer,by reducing prostaglandin E2 levels[15]. But this is not conclusive so far and the area needs more research[15].

Polyphenols (flavonoids, tannins and phenolic acids)

Sources[32,55]:

Flavanols including epicatechins: tea including green tea, cocoa and broad beans, grape seed extract, apples, onions and tea. Gallocatechins and theaflavins: tea, procyanidins from cocoa.Flavanones: citrus fruit.Hydroxycinnamic acids and phenolic acids: coffee, fruit eg. plums and pears, vegetables.
Resveratrol: skin of red grapes, peanuts, berries including mulberries, red wine.
Anthocyanins: fruit eg. purple or blue berry fruits and vegetables, grape seed proanthocyanadins (GSPs).

Polyphenols are a group of substances, including flavonoids, tannins and phenolic acids, originally found in plant-based foods and now developed in synthetic forms. In terms of skincare, flavonoids have been described as anti-inflammatory, antioxidant and UV absorbing, and may help to prevent DNA damage, cell damage and cell death, with some clinical trial evidence to support this for oral intake and topical (applied to the skin) application[40].

They are thought to act like antioxidants, depending on how much of the substance makes it across the gut and into the bloodstream, and whether there are cellular targets for it. In plants, they provide colour and protect against stresses like sunlight. While supplements deliver a suitable dose, polyphenols often work together with other nutrients to produce their effect and so it may be best to consume the polyphenol within food, in its natural state[32].

Grape seed extract (vitis vinifera) is rich in antioxidant polyphenol compounds like epicatechin galleate and proanthocyanidin, with potent antioxidant action as a result[56,57]. Grape seed proanthocyanadins (GSPs) may protect against cancerous changes from sunlight, possibly by reducing the amount of immunosuppression caused by sunlight in the skin[57].

Polypodium Leucotomos

Sources:	Supplements

Polypodium leucomotos is a fern extract, also known as Fernblock, which can act as an antioxidant because of its high concentration of phenolic compounds. It can be administered as part of a cream or as an oral supplement. It has been described as anti-inflammatory and antioxidant and may protect skin from the harmful effects of sunlight by preventing the production of ROS, reducing inflammation and preventing DNA damage and cell death[58]. It also seems to protect the antioxidant systems inside the body[58,59]. These actions may protect the skin from signs of ageing caused by the sun, sunburn and potentially the development of skin cancer[59]. Its actions may be useful in the management of several skin diseases, including polymorphic light eruption, vitiligo, melasma, psoriasis and eczema, as an add-on to existing treatments[58,59]. It has no reported side effects[58,59].

Squalene

Sources[60]:	Olive oil,wheatgerm oil, amaranth oil, rice bran oil, shark liver oil, supplements.

Squalene appears to be very important in protecting the skin from damage by free radicals. Its antioxidant action can protect the skin from the harmful effects of sunlight[60,61]. In humans, most of the squalene we absorb from our diet is distributed in our skin[60]. It is present in sebum and is also one of the most common lipids produced by human skin cells, but this production reduces after about the age of 30, contributing to the reasons for skin dryness with increasing age[61]. Squalene can be obtained through diet, probably most easily in olive oil, though its name originates from the large quantities found in shark liver[60]. Squalane is a saturated form of squalene, less susceptible to oxidation and so is more commonly used in moisturisers[60,61].

Vitamin A, Retinol (see also Carotenoids)

Sources[24,36,40]:	Liver, egg yolk, whole milk, cheese, butter, some reduced fat spreads, cod liver oil, oily fish.

Vitamin A is essential in maintaining the normal structure and function of the skin and must be obtained through the diet[29,38]. Vitamin A deficiency is common in many developing countries, especially among young children[38].

Vitamin A works as an antioxidant by neutralising free radicals, preventing ROS reactions and oxidative stress[24,29]. The vitamin A family is known as retinoids. Carotenoids can act as pro-vitamins, being converted into retinoids within the body. Once ingested, retinol (vitamin A1) from animal derived foods may carry out its functions as retinol, or it may be converted into retinaldehyde (retinal), and/or all-trans-retinoic acid (ATRA), which carries out the majority of the functions related to vitamin A within the body[39].
As a supplement, Vitamin A can have significant side effects. Too much vitamin A during pregnancy increases the risk of birth defects (teratogenicity). Pregnant women and those seeking to conceive are advised to avoid high vitamin A intake. Liver is high in vitamin A, hence part of the reason for avoiding paté[29,38]. In general, too much vitamin A can lead to acute or chronic poisoning, problems with the blood, bone pain and problems with bone growth, and can also be fatal. Children and people with liver or kidney disease are especially susceptible to side effects[29].

Synthetic Retinoids

Development of synthetic compounds related to vitamin A began in the 1960's[62]. Retinoids influence many body processes including cell growth and differentiation, cell surface alterations and immune function, and synthetic retinoid drugs are used in the management of acne, psoriasis, some genetic skin conditions and in the prevention of some forms of skin cancer[62]. Topical ATRA has been available on prescription (known as tretinoin) since around 1962 and is widely used in skincare[39]. Oral retinoid drugs can produce significant side effects, including birth defects, mood changes, changes in blood lipid profiles, liver toxicity and bone problems.

Vitamin C

Sources[8,38]:	Fresh fruits especially citrus fruits and berries, broccoli, cabbage and green leafy vegetables, peppers, tomatoes, potatoes (especially new potatoes), mangoes, papaya, banana and melon.

Vitamin C is a water soluble antioxidant which acts in the fluids inside and outside of cells[24,29]. Vitamin C is important in the production of collagen, in healing and in immune function. It reduces inflammation and acts as an antioxidant by deactivating ROS, peroxides and free radicals. It works with and regenerates vitamin E, and also increases the absorption of iron across the gut.

New collagen cannot be formed without Vitamin C, which is part of the basis for its inclusion in many anti-ageing treatments[30]. The antioxidant activity of Vitamin C alongside vitamin E protects the skin from harmful effects of sunlight, and exposure to sunlight and pollution reduce the levels of Vitamin C in the skin[30].

Vitamin C is almost side-effect free, but at large doses it can cause kidney stones, and it can increase iron absorption to harmful levels in some people. Taking large doses can lead to tolerance or dependence and then any sudden, significant reduction in intake can lead to deficiency symptoms including scurvy[29].

Vitamin D

Sources[38]:

Vitamin D2: mushrooms, fortified foods, supplements.
Vitamin D3: sunlight, fish oils, oily fish including fresh tuna,
mackerel, salmon. Liver, egg yolk, butter. Supplements.

Almost one fifth of UK adults have a low vitamin D status[28]. People at higher risk of vitamin D deficiency at northern latitudes include premature babies, children younger than 5 years, pregnant women, people older than 65 years and people with darker skin pigmentation[63,64]. At northern latitudes most of our vitamin D is produced in the skin in response to summer sunlight, but we also get some vitamin D3, and all our vitamin D2, from our diet. Vitamin D2 is used in food fortification in some countries and both vitamins D2 and D3 can be ingested as supplements.

The UK advice on vitamin D supplement intake was updated in 2016. It is now recommended that the general population aged 5 years and over should consider supplementation, especially over the winter months. People at risk of vitamin D deficiency, pregnant or breastfeeding women and children aged 1-5years should take 10micrograms (mcg) per day (400 International Units, IU) regardless of the time of year. Babies should receive 8.5-10mcg per day from birth to 1 year of age, again regardless of the time of year, unless they are consuming more than 500mls of formula per day, as this already contains enough vitamin D[65,66,67]. Vitamin D is discussed in more detail in Chapter 5.

While the nature of the chemical reactions in the skin do not allow vitamin D toxicity to develop from sunlight, very high intakes of vitamin D supplements can lead to vomiting and constipation, weight loss, weakness, kidney stones and high calcium levels[68].

Vitamin E

Sources[24,38,40]:	Vegetables and vegetable oils (soy bean, corn, safflower), spinach, avocados, corn, sunflower seeds, soy, wholegrains, nuts (especially almonds), margarine, mayonnaise, wheatgerm, some meat and dairy products, olives.

Vitamin E, which includes the tocopherol and tocotrienol compounds, is lipid soluble and lives in cell membranes. It exerts its antioxidant activity by reacting directly with free radicals, protecting cell membranes from peroxidation[24,29]. Its most biologically active form is alpha-tocopherol. Vitamin E is important for the synthesis of collagen, activation of the immune system and generation of energy. It regenerates beta-carotene and protects vitamin A from oxidation[29]. Vitamin E is thought to reduce skin dryness, particularly in people taking oral retinoids[22], and is especially abundant in the stratum corneum, where it is delivered by sebum[22,30].

Vitamins C and E often act together[30]. Vitamin E in the cell membranes is backed up by vitamin C in fluids inside and outside cells, which regenerates vitamin E as well as having its own actions. Vitamin E may provide protection from the harmful effects of sunlight when combined with other antioxidants, such as vitamin C, but neither vitamin E or C can provide similar protection when used individually[40,53]. Vitamin E levels in the skin are reduced by exposure to sunlight and pollution[30].

High doses of vitamin E can lead to problems with blood clotting, and people taking blood thinners should avoid large doses. Large doses of vitamin E may act as pro-oxidants, decrease stores of vitamin A, and increase blood glucose and lipid levels[29].

Water

The water content of the epidermis has a great impact on how our skin looks and feels. It is largely determined by the integrity of the skin barrier and the level of water loss across the epidermis, but staying hydrated will contribute to water levels in the epidermis, and is good for general health. Not drinking enough water can also contribute to the causes of a dry mouth and dry, chapped lips.

Zinc

Sources[24,36]:	High protein meats such as beef, chicken and lamb, fish and crustaceans, dairy products, cereals, seeds.

Zinc has been used as a skin treatment for centuries. Inside the body, zinc is an essential micronutrient needed for the regulation of lipid, protein and nucleic acid metabolism, normal immune function, wound repair and control of inflammation, and it also has antioxidant properties[69]. Around 6% of the body's zinc is in the skin but there is no storage mechanism for it, so continuous supply from the diet is needed[69].

Diet and Dry Skin

The integrity of the skin barrier and the hydration of the skin may be improved by including foods such as avocados, borage seed oil, evening primrose oil, fish (particularly fresh tuna, trout, mackerel and salmon), flaxseed oil, hemp seed, nuts, olive oil, olives, peanuts, soy, sunflower oil and walnuts in the diet[22]. Other elements that have been associated with hydrating the skin include glucosamine[47], nicotinamide[49], polypodium leucomotos[58], squalene[45], vitamins A[29,40] and E[22,30] and omega-3 and -6[18,21,52,54]. Biotin may improve brittle nails associated with dry skin[34].

❖

Adjusting our diet may help to improve the dryness associated with cold weather

❖

Diet and Rosacea

Dietary changes often play a role in the management of rosacea. Hot or spicy foods and alcohol are often among the triggers for flare-ups. In one study, 78% of participants with rosacea had made changes to their diet, and 95% of those had noticed a corresponding reduction in flares[70]. Medical treatments that reduce inflammation and suppress oxidative stress improve rosacea, so as well as avoiding dietary triggers, it is possible that including antioxidant and anti-inflammatory elements in the diet may have a positive effect[71,72]. There have been promising results from studies looking at the anti-inflammatory effects of omega-3 EPA and DHA, and zinc supplements[70]. Nicotinamide can also reduce inflammation[49], and dietary antioxidants which may be of benefit include vitamins[24,29], flavonoids[32,40], polypodium leucomotos[58,59], co-enzyme Q10[45], zinc[69] and grape seed extract[56,57]. There is no firm evidence of benefit from using these however.

Diet and Sun Protection

> ❖
> Dietary adjustments and supplements do not provide as much protection as sunscreen[53]
> ❖

In the skin, sunlight leads to inflammation and the release of ROS and free radicals that can consume our antioxidant systems and produce oxidative stress[25]. As explained in Chapter 1, while cold weather is not generally a time that we think about sun protection, the pattern of sun exposure that we receive at northern latitudes, with short sharp bursts during sunny days at home or sunny holidays abroad, has particular risks for the health of our skin, including sunburn, sun damage, pre-cancerous and cancerous skin lesions. Our skin is often not well prepared for strong sunlight and its defence and repair processes may become easily overwhelmed.

The best protection that we can add for our skin is to be aware of the UV index and our own skin type, wear sunscreen and protective clothing when appropriate and alter our behaviour to avoid sunburn. But, certain dietary molecules and elements may help to optimise our antioxidant and repair systems and may be a useful addition to these traditional measures, particularly for those with fair skin at high risk of sunburn. Diet or supplements should never be used as a replacement for traditional measures as they do not provide as much protection as sunscreen[53].

The use of antioxidants in the prevention and repair of UV damage has been widely studied but there is little conclusive evidence for individual molecules. Polypodium leucotomos may reduce the appearance of sunburn and help prevent sun damage and skin cancer by its antioxidant action[59]. Other antioxidants that may protect against the effects of sunlight include polyphenols such as flavonoids and grape seed extract[32,40,56,57], carotenoids (beta-carotene, lycopene, lutein)[40,43], nicotinamide[49], omega-3 fatty acids[15,16,51], vitamins C and E together[40,30,53], and multivitamins[73].

Seasonal Nutriceuticals

To help with working towards a healthy balanced diet to deliver the elements just discussed, these tables indicate foods and their vitamin, mineral and molecular benefits, along with when they are in season in the UK. The tables aren't exhaustive but can be used as a guide.

SPRING	Meat	Fruit & Vegetables	Fish	Other
Vitamin A				
Retinol	Liver			Whole milk, cheese, butter, spreads
Carotenoids		Carrots, dark green leafy vegetables		
Vitamin C		Gooseberries, rhubarb , green leafy vegetables eg. spinach, broccoli. Cauliflower, asparagus, cabbage, spring onion		
Vitamin E		Spinach, asparagus		Nuts
Other antioxidants & phytochemicals		Gooseberries, rhubarb, curly kale, spinach, Savoy cabbage, broccoli		
B Vitamins	Beef, spring lamb	Asparagus, celeriac		
CoEnzyme Q10		Spinach	Crab, haddock, John Dory, lobster, mackerel, monkfish, prawns, sea bass, salmon, trout, turbot	
Omega-3 Fatty Acids	Grass fed lamb		Mackerel, trout, salmon, haddock, crab	Omega-3 eggs, nuts
Iron	Beef, spring lamb	Asparagus, celeriac, cabbage		
Zinc	Beef			

References: Season[74] Nutrients[33,35,36,37,55]

SUMMER	Meat	Fruit & Vegetables	Fish	Other
Vitamin A				
Retinol	Liver			Whole milk, cheese, butter, spreads
Carotenoids		Carrots, courgette, peas, tomatoes, squash, dark green leafy vegetables, green beans		
Vitamin C		Greengages, loganberries, Tay berries Cauliflower, carrots, courgette, fennel, peas, squash, garlic, potatoes, radishes, green beans, beetroot		
Vitamin E				Nuts
Other antioxidants & phytochemicals		Blueberries, currants, strawberries, raspberries, plums, courgette, potatoes, beetroot, garlic, tomatoes, broad beans		
B Vitamins, Biotin	Beef, lamb, pork, venison	Carrots, potatoes, garlic		
CoEnzyme Q10			Crab, pilchards, wild salmon	
Omega-3 Fatty Acids	Grass fed lamb		Crab	Omega-3 eggs, nuts
Iron	Beef, lamb, pork, venison	Broad beans, garlic		
Zinc	Beef, pork, venison	Broad beans, garlic		

References: Season[74] Nutrients[33,35,36,37,55]

AUTUMN	Meat	Fruit & Vegetables	Fish	Other
Vitamin A				
Retinol	Liver			Whole milk, cheese, butter, spreads
Carotenoids		Elderberries, squash, pumpkin, dark green leafy vegetables, watercress, lettuce, rocket		
Vitamin C		Pears, elderberries, damsons Potatoes, sweetcorn, rocket, watercress		
Vitamin E				Nuts
Other antioxidants and phytochemicals		Blackberries, plums, pears, elderberries, apples potatoes		
B Vitamins and Biotin	Lamb, grouse, ham, pork, venison	Elderberries, potatoes, mushrooms		
CoEnzyme Q10			Brill, dab, Dover sole, flounders, oysters, skate	
Omega-3 Fatty Acids	Heather fed lamb			Omega-3 eggs, nuts
Iron	Lamb, pork, venison			
Zinc	Pork, ham, venison			

References: Season[74] Nutrients[33,35,36,37,55]

WINTER	Meat	Fruit & Vegetables	Fish	Other
Vitamin A				
Retinol	Liver			Whole milk, cheese, butter, spreads
Carotenoids		Carrots, dark green leafy vegetables, bay leaves, Brussels sprouts, leek		
Vitamin C	Goose	Pears, quince, turnips, Brussels sprouts, cauliflower, bay leaves, fennel, leek, potato, celeriac, cabbage		
Vitamin E				Nuts
Other antioxidants and phytochemicals		Pears, quince, apples Curly kale, Brussels sprouts, parsnip, cabbage, fennel, potatoes		
B Vitamins and Biotin	Venison, duck, turkey, ham, pheasant, goose	Leek, celeriac, bay leaves		
CoEnzyme Q10			Grey mullet, mussels, scallops	
Omega-3 Fatty Acids				Omega-3 eggs, nuts
Iron	Venison, turkey, duck, goose, pheasant	Brussels sprouts, leek, bay leaves, celeriac, cabbage		
Zinc	Venison, turkey, duck, goose, pheasant, ham	Bay leaves		

References: Season[74] Nutrients[33,35,36,37,55]

Summary

- Cold weather can encourage us to eat less healthy food, which can have an impact on our skin.
- Eating lots of sugar can affect skin structure and function through the production of advanced glycation products (AGEs).
- Eating too much fat is bad for us, but essential fatty acids, including polyunsaturated fatty acids (PUFAs) and cholesterol are important components of the skin barrier.
- Adjusting our diet may improve our health and help limit the effects of cold weather on our skin by maintaining our antioxidant and anti-inflammatory systems.
- No one magic ingredient will resolve skin problems, and overdosing on any one element can be as dangerous as deficiency.
- Any dietary alterations should complement established or prescribed medical treatment, not used in place of it.

References

1. The Department for Environment, Food and Rural Affairs. The UK's food history revealed through five generations of data [Internet]. United Kingdom: www.gov.uk; [updated 2016 Sept.; cited 2019 June]. Available from: https://www.gov.uk/government/news/the-uks-food-history-revealed-through-five-generations-of-data
2. Devlin S. Agricultural Labour in the UK. Food research collaboration policy brief. 2016 July. Available online from: https://www.sustainweb.org/resources/files/reports/Agricultural_Labour_InThe_UK.pdf
3. Department for Environment Food and Rural Affairs. Food statistics pocketbook 2016. 2016 Sept. Available online from: https://assets.publishing.service.gov.uk/government/uploads/system/uploads/attachment_data/file/608426/foodpocketbook-2016report-rev-12apr17.pdf
4. Department for Environment Food and Rural Affairs. Food statistics in your pocket 2017 – global and UK supply. 2018 Oct. Available online from: https://www.gov.uk/government/publications/food-statistics-pocketbook-2017/food-statistics-in-your-pocket-2017-global-and-uk-supply
5. University of Cambridge Research News. Price Gap between more and less healthy foods grows [Internet]. United Kingdom: University of Cambridge; [updated 2014 Oct; cited 2019 June]. Available from: www.cam.ac.uk/research/news/price-gap-between-more-and-less-healthy-foods-grows
6. University of Oxford, The Food Foundation, Sustain the Alliance for Better Food and Farming. Measuring household food insecurity in the UK and why we must do it [Internet]. 2016 [cited 2019 June]. Available from: https://foodfoundation.org.uk/wp-content/uploads/2016/07/MeasuringHouseholdFoodInsecurity.pdf
7. Hines S. The hidden sugars in your favourite Christmas foods, and what you can replace them with [Internet]. United Kingdom: Goodhousekeeping.com; [updated 2017 Dec.; cited 2019 June]. Available from: https://www.goodhousekeeping.com/uk/health/healthy-eating/a574227/low-sugar-christmas-food/

8. Danby FW. Nutrition and aging skin: sugar and glycation. Clin Dermatol. 2010;28(4):409-411
9. National Health Service (NHS). How much sugar is good for me [Internet]. United Kingdom: NHS; [2018 May; cited 2019 June]. Available from: https://www.nhs.uk/common-health-questions/food-and-diet/how-much-sugar-is-good-for-me
10 . National Health Service (NHS). How does sugar in our diet affect our health [Internet]. United Kingdom: NHS; [2017 Aug; cited 2019 June]. Available from: www.nhs.uk/live-well/eat-well/how-does-sugar-in-our-diet-affect-our-health
11. Nguyen HP, Katta R. Sugar sag: glycation and the role of diet in aging skin. Skin Therapy Lett. 2015;20(6):1-5
12. Davis BC, Kris-Etherton PM. Achieving optimal essential fatty acid status in vegetarians: current knowledge and practical implications. Am J Clin Nutr. 2003;78(3):640S-646S
13. National Health Service (NHS). Fat: the Facts [Internet]. United Kingdom: NHS; [2017 May; cited 2019 June]. Available from: www.nhs.uk/live-well/eat-well/different-fats-nutrition
14. Vieira S, McClements DJ, Decker EA. Challenges of utilising healthy fats in foods. Adv Nutr. 2015;6(3):309S-317S
15. Black HS, Rhodes LE. Potential benefits of omega-3 fatty acids in non-melanoma skin cancer. J Clin Med. 2016;5(2):23
16. Puglia C, Tropea S, Rizza L, Santagati NA, Bonina F. In vitro percutaneous absorption studies and in vivo evaluation of anti-inflammatory activity of essential fatty acids (EFA) from fish oil extracts. Int J Pharm. 2005;299(1-2):41-48
17. Huang T, Wang PW, Yang SC, Chou WL, Fang JY. Cosmetic and therapeutic applications of fish oil's fatty acids on the skin. Mar Drugs. 2018;16(8). Pii:E256
18. Burton JL. Dietary fatty acids and inflammatory skin disease. Lancet. 1989;1(8628):27-31
19. Loden M. The clinical benefit of moisturizers. J Eur Acad Dermatol Venereol. 2005;19(6):672-688
20. Key TJ, Appleby PN, Rosell MS. Health effects of vegetarian and vegan Diets. Proc Nutr Soc. 2006;65(1):35-41

21. Brosche T, Platt D. Effect of borage oil consumption on fatty acid metabolism, transepidermal water loss and skin parameters in elderly people. Arch Gerontol Geriatr. 2000;30(2):139-150
22. Baumann L, Saghari S, Weisberg E. Cosmetic dermatology: principles and practice. 2nd ed. United States of America: McGraw-Hill Companies Inc.; 2009
23. National Health Service (NHS). High Cholesterol [Internet]. United Kingdom: NHS; [2019 Apr.; cited 2019 June]. Available from: www.nhs.uk/conditions/high-cholesterol
24. Rauma AL, Mykkanen H. Antioxidant status in vegetarians versus omnivores. Nutrition 2000;16(2):111-119
25. Addor FA. Antioxidants in dermatology. An Bras Dermatol. 2017;92(3):356-362
26. Lenaz G. Role of mitochondria in oxidative stress and ageing. Biochim Biophys Acta. 1998;1366(1-2):53-67
27. Halvorsen BL, Holte K, Myhrstad MC, Barikmo I, Hvattum E, Remberg SF, Wold AB, Haffner K, Baugerod H, Andersen LF, Moskaug O, Jacobs DR Jr, Blomhoff R. A systematic screening of total antioxidants in dietary plants. J Nutr. 2002;132(3):461-471
28. Public Health England, Food Standards Agency. The national diet and nutrition survey: results from years 1,2,3 and 4 (combined) of the rolling programme (2008/2009 – 2011/2012). 2017 Feb. Available online from: https://assets.publishing.service.gov.uk/government/uploads/system/uploads/attachment_data/file/594360/NDNS_Y1_to_4_UK_report_executive_summary_revised_February_2017.pdf
29. Rutkowski M, Grzegorczyk K. Adverse effects of antioxidative vitamins. Int J Occup Med Environ Health. 2012;25(2):105-121
30. Burke KE. Interaction of vitamins C and E as better cosmeceuticals. Dermatol Ther. 2007;20(5):314-321
31. Schagen SK, Zampelli VA, Makrantonaki E, Zouboulis CC. Discovering the link between nutrition and skin aging. Dermatoendocrinol. 2012;4(3):298-307
32. Williamson G. The role of polyphenols in modern nutrition. Nutr Bull. 2017;42(3):226-235
33. Watson W, Butler N. Biotin-rich foods [Internet]. United

States of America: Healthline Media; [2017 May; cited 2019 Jun.]. Available from: https://www.healthline.com/health/biotin-rich-foods

34. Cashman MW, Sloan SB. Nutrition and nail disease. Clin Dermatol. 2010;28(4):420-425

35. Harvard T.H. Chan School of Public Health. The Nutrition Source [Internet]. United States of America: Harvard TH Chan School of Public Health; [unknown; cited 2019 June].Available from: https://www.hsph.harvard.edu/nutritionsource/food-features

36. NutritionData.Self.com. Nutrients in food tool [Internet]. United States of America: Conde Nast; [2014; cited 2019 June]. Available from: https://nutritiondata.self.com/tools/nutrient-search

37. NutritionValue.org. Find nutritional value of a product [Internet]. United States of America: NutritionValue.org; [2019; cited 2019 June]. Available from: https://nutritionvalue.org

38. British Nutrition Foundation. Nutrients, food and ingredients. [Internet]. United Kingdom: British Nutrition Foundation; [updated unknown; cited 2019 June]. Available from: https://www.nutrition.org.uk/nutritionscience/nutrients-food-and-ingredients.html

39. Blomhoff R, Blomhoff HK. Overview of retinoid metabolism and function. J Neurobiol. 2006;66(7):606-630

40. Evans JA, Johnson EJ. The role of phytonutrients in skin health. Nutrients. 2010;2(8):903-928

41. Huang HY, Caballero B, Chang S, Alberg A, Semba R, Schneyer C, Wilson RF, Cheng TY, Prokopowicz G, Barnes GJ 2nd, Vassy J, Bass EB. Multivitamin/mineral supplements and prevention of chronic disease. Evid Rep Technol Assess (Full Rep). 2006;(139):1-117

42. Tanvetyanon T, Bepler G. Beta-carotene in multivitamins and the possible risk of lung cancer among smokers versus former smokers: a meta-analysis and evaluation of national brands. Cancer 2008;113(1):150-157

43. Wright TI, Spencer JM, Flowers FP. Chemoprevention of nonmelanoma skin cancer. J Am Acad Dermatol. 2006;54(6):933-946

44. Pravst I, Zmitek K, Zmitek J. Coenzyme Q10 contents in foods and fortification strategies. Crit Rev Food Sci Nutr. 2010;50(4):269-280

45. Beyer RE. The participation of coenzyme Q in free radical production and antioxidation. Free Radic Biol Med. 1990;8(6):545-565
46. Fuller B, Smith D, Howerton A, Kern D. Anti-inflammatory effects of CoQ10 and colourless carotenoids. J Cosmet Dermatol. 2006;5(1):30-38
47. Bissett DL. Glucosamine: an ingredient with skin and other benefits. J Cosmet Dermatol. 2006;5(4):309-315
48. Kikuchi K, Matahira Y. Oral N-Acetylglucosamine supplementation improves skin conditions of female volunteeers: clinical evaluation by microscopic three dimensional skin surface analyser. J Appl Cosmetol. 2002;20:143-152
49. Wohlrab J, Kreft D. Niacinamide – mechanisms of action and its topical use in dermatology. Skin Pharmacol Physiol. 2014;27(6):311-315
50. Ranaweera A. Nicotinamide [Internet]. New Zealand: DermNetNZ.org; [2018 March; cited 2019 June]. Available from: https://www.dermnetnz.org/topics/nicotinamide
51. Sies H, Stahl W. Nutritional protection against skin damage from sunlight. Annu Rev Nutr. 2004;24:173-200
52. Bjorneboe A, Soyland E, Bjorneboe GE, Rajka G, Drevon CA. Effect of dietary supplementation with eicosapentaenoic acid in the treatment of atopic dermatitis. Br J Dermatol. 1987;117(4):463-469
53. Boelsma E, Hendriks HF, Roza L. Nutritional skin care: health effects of micronutrients and fatty acids. Am J Clin Nutr. 2001;73(5):853-864
54. Schalin-Karrila M, Mattila L, Jansen CT, Uotila P. Evening primrose oil in the treatment of atopic eczema: effect on clinical status, plasma phospholipid fatty acids and circulating blood prostaglandins. Br J Dermatol. 1987;117(1):11-19
55. Phytochemicals.info. Phytochemicals [Internet]. Belgium: Phytochemicals.info; [unknown; cited June 2019]. Available from: https://www.phytochemicals.info
56. Hemmati AA, Foroozan M, Houshmand G, Moosavi ZB, Bahadoram M, Maram NS. The topical effect of grape seed extract 2% cream on surgery wound healing. Glob J Health Sci. 2014;7(3):52-58
57. Katiyar SK. Grape seed proanthocyanidines and skin cancer prevention: inhibition of oxidative stress and

protection of the immune system. Mol Nutr Food Res. 2008;52(1):S71-76

58. Choudry SZ, Bhatia N, Ceilley R, Hougeir F, Lieberman R, Hamzavi I, Lim HW. Role of oral Polypodium leucomotos extract in dermatologic diseases: a review of the literature. Drugs Dermatol. 2014;13(2):148-153

59. Gonzalez S, Gilaberte Y, Philips N, Juarranz A. Fernblock, a nutriceutical with photoprotective properties and potential preventive agent for skin photoaging and photoinduced skin cancers. In J Mol Sci. 2011;12(12):8466-8475

60. Kelly GS. Squalene and its potential clinical uses. Altern Med Rev. 1999;4(1):29-36

61. Sethi A, Kaur T, Malhotra SK, Gambhir ML. Moisturizers: the slippery road. Indian J Dermatol. 2016;61(3):279-287

62. Mukherjee S, Date A, Patravale V, Korting HC, Roeder A, Weindl G. Retinoids in the treatment of skin aging: an overview of clinical efficacy and safety. Clin Interview Aging. 2006;1(4):327-348

63. Lips P, van Schoor NM, de Jongh RT. Diet, sun and lifestyle as determinants of vitamin D status. Ann N Y Acad Sci. 2014;1317:92-98

64. Madsen KH, Rasmussen LB, Mejborn H, Andersen EW, Molgaard C, Nissen J, Tetens I, Andersen R. Vitamin D status and its determinants in children and adults among families in late summer in Denmark. Br J Nutr. 2014;112(5):776-784

65. Scientific Advisory Committee on Nutrition (SACN). Vitamin D and Health. 2016 July. Available online from: https://assets.publishing.service.gov.uk/government/uploads/system/uploads/attachment_data/file/537616/SACN_vitamin_D_and_health_report.pdf

66. National Health Service (NHS). Vitamin D [Internet]. United Kingdom: NHS; [2017 Mar.; cited 2019 June]. Available from: https://www.nhs.uk/conditions/vitamins-and-minerals/vitamin-d

67. Calderwood C. Correspondence: new recommendations on vitamin D supplementation. Chief Medical Officer Directorate. 2017 Nov. Available online from: https://www.cps.scot/media/1987/cmo-unnumbered-letter-issued-on-24-november-2017-final-new-recomme.pdf

68. Laird E, Ward M, McSorley E, Strain JJ, Wallace J. Vitamin D and bone health: potential mechanisms. Nutrients. 2010;2(7):693-724
69. Gupta M, Mahajan VK, Mehta KS, Chauhan PS. Zinc in dermatology: a review. Dermatol Res Pract. 2014;2014:709152
70. Weiss E, Katta R. Diet and rosacea: the role of dietary change in the management of rosacea. Dermatol Pract Concept. 2017;7(4):31-37
71. Levin J, Miller R. A guide to the ingredients and potential benefits of over-the-counter cleansers and moisturizers for rosacea patients. J Clin Aesthet Dermatol. 2011;4(8):31-49
72. Yamasaki K, Gallo RL. The molecular pathology of rosacea. J Dermatol Sci. 2009;55(2):77-81
73. Asgari MM, Chren MM, Warton EM, Friedman GD, White E. Supplement use and risk of cutaneous squamous cell carcinoma. J Am Acad Dermatol. 2011;65(6):1145-1151
74. LoveBritishFood.co.uk. What's in season when [Internet]. United Kingdom: LoveBritishFood.co.uk; [2019; cited 2019 June]. Available from: http://www.lovebritishfood.co.uk/british-food-and-drinks/whats-in-season-when

5. Bring me Sunshine: Occasional Sun and the Skin

5. Bring Me Sunshine:
Occasional Sun and the Skin

At northern latitudes many of us look forward to clear blue skies, dry weather and warmth. After spending autumn and winter wrapped up in layers of clothing we can enjoy being outdoors, doing things that make us happy and improve our health. Sunlight and exercise both raise our levels of beta-endorphin, a hormone that creates a feeling of wellbeing[1]. But in northern parts of the world we tend to experience sunlight in short, sharp bursts, during a nice few weeks at home or on sunny holidays abroad, which has consequences for our skin and our general health.

Sunshine, skin and health have a complicated relationship that centres on balance. On reaching our skin, ultraviolet B (UVB) rays start the production of vitamin D, which is important for our bones and probably other areas of our health. UV light also triggers the production of nitric oxide in our skin, which can reduce our blood pressure, with knock-on benefits like lowering our risk of stroke and coronary heart disease[2]. On the other hand, too much sun exposure can lead to sun damage and skin cancer.

Sunshine, skin and health have a complicated relationship that centres on balance.

The right balance between sun exposure and sun protection varies between each of us, depending on our personal characteristics and how we live our lives. Some people are more at risk of sun damage and skin cancer while others are more at risk of vitamin D deficiency. This chapter explores our relationship with sunlight and how we can each find our own balance.

Sunshine and The Solar Spectrum

The sun emits all kinds of radiation, but by the time its rays reach the Earth only visible light, ultraviolet (UV) and infrared rays are left. Together these rays are known as the solar spectrum and form our sunlight[3]. Within the spectrum, we are generally only able to see visible light, which is the fraction that can be divided into the rainbow. UV rays have important effects on our skin and cell DNA, while infrared rays contribute significantly to the warmth provided by the sun.

UV is divided into three wavelength ranges: UVA, UVB and UVC. More than 95% of the suns rays reaching the Earth are UVA. UVC is absorbed in the Earth's atmosphere and does not reach the surface. At the latitudes of the UK, from 49.5°North in the Isles of Scilly to 60.5°North in Shetland, UVA reaches us all year round, but UVB is seasonal, reaching us between roughly April and September[4,5,6]. UV rays are most intense when the sun is high in the sky, between 11am and 3pm[7]. Only UVB stimulates molecules in our skin to make vitamin D.

Sun and the Skin

A Brief History

Genetically pale skinned Northern Europeans have chased after a suntan for almost 100 years but things used to be very different. Many Victorian women for example cultivated an invalid-like pallor that was the fashion at the time and a sign that their status in society meant they never had to work outside in the sun[8].

For the Victorians, upkeep of pale skin demanded the use of hats and parasols, as are still in use today, but also interesting remedies such as arsenic. Arsenic causes a number of harmful effects inside the body, after all it can also be used as a poison, but it does reduce pigmentation in the skin. Although it was known to be toxic, arsenic was used for decades because it was effective and felt to be more 'natural' than painting the skin with white powders or enamels[8]. This is an excellent example of why, although something may be labelled 'natural', that doesn't automatically mean that it is good for us.

Although something may be labelled 'natural' that doesn't automatically mean that it is good for us.

Attitudes towards the suntan changed dramatically with the Industrial Revolution, which moved many poorer workers from outdoor to indoor settings with long working hours and very little leisure time, to the detriment of their suntan and obviously other aspects of their health[9]. Cultural changes during the 19th and 20th centuries saw wealthy people with time to spare spending more and more time outdoors, taking part in sports and visiting outdoor recreation facilities, and by the 1920's having a tan had become a sign of leisure, travel and wealth, commented on in the novels of F. Scott Fitzgerald and modelled by fashion icons like Coco Chanel[9].

❖
The term 'sunlight cancer' was first used around 1933[9]
❖

Important medical discoveries also backed the benefits of sunlight. In the 1890's TA Palm showed that lack of sunlight was the chief cause of rickets and in the early 1900's physicians began to treat children with artificial UV light for rickets and other conditions including tuberculosis[9,10,11]. By 1910, medical journals were describing suntanned skin as 'an index of health' and 'as important as a water bath' in terms of children's health[9,12].

But, the term 'sunlight cancer' was first used around 1933 and later in the 1940's medical and public awareness of the dangers of excessive UV exposure was on the rise[9]. Gradually, UVB was established as the cause of sunburn, UVB induced DNA mutations were identified and UVB was established as a carcinogen. In the 1980's problems with the ozone layer were identified, leading to an increase in media articles reporting the harmful effects of UV and an increase in public health campaigns on safe sun exposure[13].

The development of cancer databases provided evidence for associations between many cancers and their causative factors. While diagnostic and reporting facilities have also improved, it is clear that cases of melanoma have increased in parallel with changes in fashion, travel, leisure activities and increased exposure to UV[9]. However, like the Victorian ladies who continued with arsenic despite knowing the harm they were causing themselves, many of us continue to bake, burn and fry ourselves in the pursuit of a suntan.

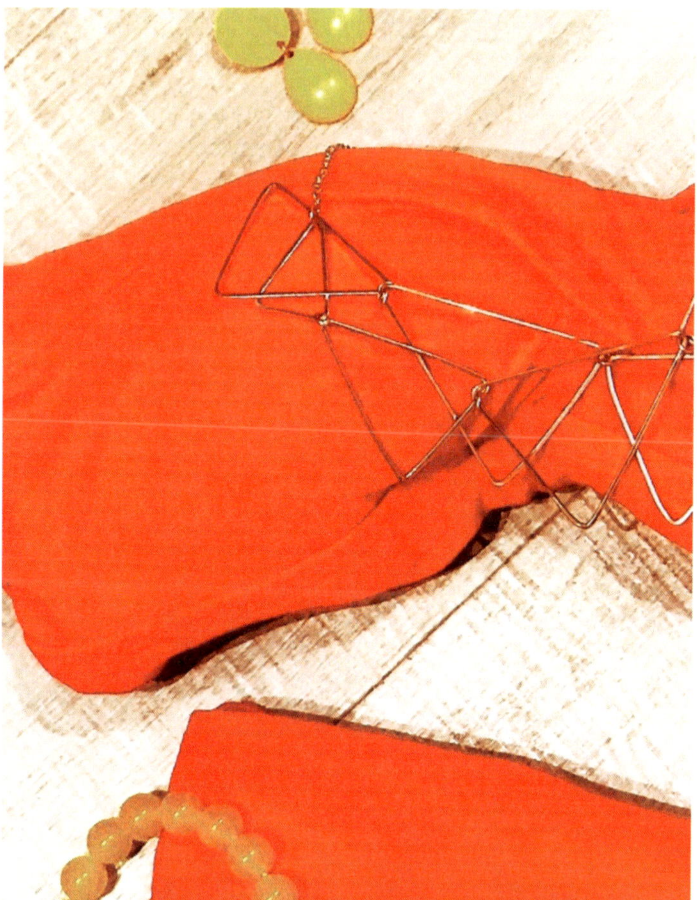

The Effect of UV Light

UV light can lead to inflammation and suppression of immune activity in the skin, with exposure generating inflammatory cell messengers called cytokines, reactive oxygen species (ROS) and free radicals. ROS and free radicals can consume our antioxidant systems resulting in a state of oxidative stress, which causes damage to cell proteins and DNA[1,14]. On reaching DNA, UVB light creates lesions called pyrimidine dimers, which contribute to the suppression of the skin's immune system[1,11]. While some damage may be recognised and repaired, the reduction in immune activity may mean that some damage is not recognised and can be replicated into new cells, leading eventually to cancerous changes developing[11].

❖

Melanoma has increased with changes in fashion, travel, leisure activities and increased skin exposure to UV[9]

❖

As discussed in Chapter 1, we have defence systems to protect the DNA in our cells from the harmful effects of UV light. Keratinocytes at the stratum corneum can reflect, scatter and absorb UV and our epidermis thickens in response to persistent sun exposure. Melanin, produced by melanocytes and packaged into melanosomes, positions itself above keratinocyte cell nuclei to protect DNA by absorbing any UV light that reaches it.

We have processes for DNA repair, processes that stop damaged cells from growing or creating new versions of themselves and processes that encourage damaged cells to die rather than reproduce. Scattered within cells, and in the fluids between them, we also have systems of antioxidants and chelating minerals working tirelessly to reduce inflammation[1]. But despite all these defences, many of us will have been sunburnt at some point in our lives and sun damage and skin cancer continue to be common, showing that our innate processes are sometimes not able to cope with the stress put on them.

> ❖
> Not getting enough non-burning sun exposure has been related to risks to health
> ❖

Taking a regular vitamin D supplement is sometimes mooted as a replacement for regular exposure to sunlight, but UV light does also have helpful effects on the skin. There are benefits to be had from regular non-burning sun exposure, both from vitamin D production and also from factors independent of vitamin D. Not getting enough non-burning sun exposure has been related to risks to general health including a higher risk of some cancers, high blood pressure and cardiovascular disease, type 2 diabetes, the metabolic syndrome and obesity, Alzheimer's disease, multiple sclerosis, type 1 diabetes, rheumatoid arthritis, psoriasis and eye problems[1]. For the benefits that are independent of vitamin D, taking a regular supplement will obviously not be an adequate substitute[1].

Suntan and Sunburn

Immediate skin pigmentation (tanning) is caused by redistribution and oxidation of melanin in the epidermis in response to UVA. Delayed tanning, peaking around 3 days after exposure, is caused by melanocytes adapting, increasing in number and producing more melanin and is a response to UVB. A tan from UVB exposure provides 5-10 times more protection against sunburn than a tan from UVA[11].

❖
Living at northern latitude is associated with a greater risk of sunburn
❖

Everyone should avoid sunburn, which is associated with substantially increased risk of skin cancer[1]. Unfortunately, living at northern latitude is associated with a greater risk of sunburn. This is not because we live in a particularly sunny place, but because there are more fair skinned people in the population and sunburn is linked to a pattern of intermittent sun exposure, which fits with our changeable climate and holiday patterns[17]. Intermittent, intense sun exposure leading to sunburn is linked to a higher risk of developing melanoma and other forms of skin cancer[1,18].

Most sunburn is caused by UVB and reaches a peak 6-24 hours after exposure. While UVB is only present in UK sunlight during the late spring and summer months of the year[4,5,6], UVA is present all year round and causes cell changes particularly affecting melanocytes and fibroblasts and is associated with early and exaggerated signs of ageing, known as photoageing. Both UVB and UVA have been linked to causing skin cancer[11].

While it is possible for skin of any colour to burn in the sun, sunburn is more of a problem for those with lighter skin types (Fitzpatrick phototypes 1-3), particularly at the latitude of the UK. A recent study including a population of women with black skin living in the UK reported that 83% of these women had never, or only rarely been sunburnt. It is also worthwhile noting that a UV index of 10 on white skin has been reported as equivalent to a UV index of around 1 on black skin, well below the level at which sun protection is usually recommended[19].

The UV Index and The Shadow Rule

The UV index was developed by the World Health Organisation (WHO) to explain the strength of UV light so that steps could be taken to protect the skin from sunburn. The intensity of UV varies with latitude, time of day and year, weather factors and altitude. UV intensity and UVB are greatest when the sun is high in the sky[7]. The UV index goes from 1 to 11+.

1-2 = Low
3-5 = Moderate
6-7 = High
8-10 = Very high
11 = Extreme

In the UK the UV index is usually 6 or below and does not go higher than 8, but indices of 9 or 10 are common towards the Mediterranean[15]. Sun protection is generally recommended when the UV index is greater than 3. Judging the length of your shadow compared to your height also gives an idea of your local UV intensity. UV rays are strong when your shadow is shorter than you[16].

Sun Damage and Skin Cancer

While our skin ages chronologically with the rest of our body, lots of sun exposure (even intermittently) can produce wrinkling, skin that is coarse and leathery in texture, patchy pigmentation and higher numbers of skin lesions. These signs are collectively known as sun damage, photodamage, or photoageing. Skin lesions may be benign growths, pre-cancerous or cancerous in nature.

There are several different forms of skin cancer but the three most well known are basal cell carcinoma (BCC), squamous cell carcinoma (SCC) and melanoma. Skin cancers are often described as melanoma or non-melanoma skin cancers, with non-melanoma skin cancer including BCC and SCC. Many forms of skin cancer are treatable if diagnosed early and removed, but if the signs of skin cancer are not recognised tumours may become large and disfiguring, or spread to distant areas of the body, leading to serious health problems and even death.

Basal Cell Carcinoma (BCC)

BCCs are a very common form of cancer. They grow very slowly, often over several years, and do not tend to spread to other places in the body. They can appear as soft or slightly firm skin coloured or pink nodules, often with visible blood vessels. Sometimes the centre may break down and ulcerate.

They can be cured by surgical removal, and if that is complete they do not need any other treatment. If they are left in place they can become disfiguring as they continue to grow, ulcerate and bleed. They can invade into nearby bone, cartilage and other tissues, making them more and more difficult to completely remove. If you notice a persisting, slowly growing lump on your skin, ask your doctor to have a look at it.

Squamous Cell Carcinoma (SCC)

SCCs are also common. They grow more quickly, usually over a few months, and can potentially spread to lymph nodes. They may appear as a red to pink scabby nodule on the skin, which often breaks down and ulcerates.
SCCs can also be cured by complete surgical removal, but once the lesion has been removed and analysed in the laboratory some of its features will give more information about the risk of it having spread, and whether any further investigation or treatment is needed. Again, if you are concerned about any persisting or growing lesions on your skin, ask your doctor to have a look.

Melanoma

Melanoma is one of the most potentially dangerous forms of skin cancer. Cases of melanoma have increased steadily during the last century[1]. Melanoma develops after melanocyte cell DNA is irreparably damaged, and appears as a change in a mole, or a changing new mole. This is why skin cancer awareness campaigns advise people to monitor their moles for changes.

Some changes in moles are normal, for example after a sunny holiday moles can appear slightly darker and freckles can appear, or a mole may bleed and appear inflamed after being rubbed on clothing or scratched at. Also, over several years, moles that were originally flat can become elevated and even skin tag like. But, if any changes are continuing, occurring quickly or worrying you please see your doctor and ask them to have a look.

Individually, our risk of melanoma is affected by our genetics, our total number of moles, our number of irregularly shaped or coloured moles and our pattern of sun exposure. Pattern of sun exposure is important, as intermittent, intense exposure to sunlight increases the risk of melanoma, which is important to be aware of given the features of our climate[17]. There is some evidence that tanning, particularly at a young age, increases the risk of melanoma, while sunburns appear to be equally risky at any age[1,20].

Sunburn has been associated with a doubling of melanoma risk, but chronic non-burning sun exposure and outdoor jobs have actually been associated with a reduced risk[1]. High dose exposure to the sun after a prolonged period of sun avoidance, such as a sunny holiday at the end of winter, is likely to cause substantial damage to melanocyte DNA, as the defence processes in our skin will have decreased with lack of exposure, and with little time for them to adapt, the skin will not be physiologically well protected[17].

Skin Monitoring

The ABCDE rule for monitoring moles has been used in public health campaigns for many years (Asymmetry; Border irregularity; uneven Colour with different shades of black, brown or pink; Diameter, which has varied from 6mm-1cm); and Evolving. This advice should be used as a guide, and should always be accompanied by an emphasis on the importance of evolving changes in lesions, regardless of whether the change in the mole fits within the other criteria.

> ❖
> Have a look at your skin every 4-6 weeks. Get to know your skin and where your moles are.
> ❖

Many people have benign but asymmetric, irregular moles that are not changing and need nothing more than being kept an eye on. On the other hand, nodular melanomas can appear as symmetrical, uniformly coloured nodules within a mole than may be less than 6mm-1cm in size. Please don't ignore moles that are changing, even if they fit within the ABCD guidance. See your doctor if you have concerns.

Have a look over your skin every 4-6 weeks. The more regularly you check, the more you will get to know your skin and where your moles are. Everyone has a back, and backs are hard to see, but it is possible to see your back by angling a mirror in front of you and one behind you. Some shop changing rooms have handily angled mirrors, and always take the chance to have a look at the moles on your back if you happen to be in one. Scalps and tops of ears are the most impossible areas to see by yourself. Helpfully, hairdressers, beauty therapists and allied health professionals are more and more aware of the importance of skin lesions and may point out any that they spot to you.

If you are not sure whether a mole is changing, put a ruler or scalable object next to it and take a dated photograph with your camera or smartphone. Look at it again every 1-2 weeks, and if it is changing compared with your picture, see your doctor. Photographs like these are useful for remembering what moles looked like at one point in time and can be used as a reference in the future. If you have very many moles then mole mapping may be useful for you. This may be done using either old-fashioned photographs or digital computer software to keep track of changes in moles. It may be available in some dermatology departments, depending on resources, or in the private sector.

Sunscreen

Sunscreens may include chemical filters (known as organic agents) to absorb UV light, and/or physical or mineral blockers to reflect it. Most contain a combination of these agents to protect against both UVB and UVA. A good sunscreen should offer broad-spectrum cover (UVB and UVA) ideally in equal balance. Some sunscreens do not filter UVA as well as they do UVB and frequent use of these can lead to unintended high UVA exposure.

Chemical filters, which absorb UV light to prevent it from reaching skin cells and their DNA, include padimate O (a PABA ester), octylmethoxycinnamate, octisalate, homosalate, trolamine salicylate, benzophenone-3, benzophenone-4, oxybenzone, avobenzone, enzacamene and octocrylene[11,21,22,23]. Chemical filters may cause skin irritation or contact allergy in sensitive people. These chemicals can be absorbed across the skin into the body and have been detected in breastmilk and placental tissue[21,22,23]. It is not yet clear if there are any real health risks from this and results from further research are awaited. Chemical filters can also accumulate in bodies of water, including swimming pools, rivers and lakes, and have been found in fish, crustaceans and other aquatic life[23].

Physical blockers (inorganic sunscreens) contain the minerals titanium dioxide or zinc oxide to reflect UV light away from the skin. They have been used from the early days of sunscreen production but due to the size of the mineral particles and the nature of reflection, preparations are quite thick and opaque in appearance. The large particle size means that the agents are not well absorbed into the skin[23], but while new technology has allowed the particles to become smaller and more cosmetically appealing, this does increase the risk of them becoming able to penetrate the skin[11,23]. The risk from exposure to zinc and titanium is felt to be low but again more research is needed[23].

There is a reality gap between the amounts of sunscreen used in product testing to determine SPF and the amounts the average human applies for daily use. To achieve the SPF on the bottle, an average adult would need to apply at least 35mls of sunscreen (roughly two tablespoons)[24]. This means that the average 200ml bottle of sunscreen would last just less than 6 applications. Using a higher SPF can help combat the reality that most people apply far less than this. The National Institute of Clinical Excellence (NICE) advise that if a person plans to be out in the sun long enough to risk burning, sunscreen should be applied 30minutes before going out, then again at the time of leaving[24].

Use of sunscreen means that the skin may take longer to acclimatise to strong UV light, but this gradual exposure is often needed in order to avoid significant damage, for example at the start of a sunny holiday[4]. Use of high factor sunscreen may affect vitamin D production in the skin, but production does not stop production completely and in some circumstances it may not have an impact on vitamin D levels at all[4,25,26,27,28]. A recent Australian review of the impact of sunscreen on vitamin D levels found that there is little scientific evidence that realistically applied sunscreen providing an SPF of roughly 16 affects vitamin D production[27]. A further recent British study investigated the impact that sunscreen use had on vitamin D during a one week long holiday in Tenerife (28°North). They found that sunscreen providing an SPF of 15 produced no significant impact on vitamin D levels when UV levels were very high[28].

Although this research demonstrates that vitamin D can still be produced while using moderate SPFs in response to high UV indices, an important issue is the impact of daily use of high factor sunscreen when the UV index in low, as is advised in many sun protection protocols for young children in the UK. This is of particular importance to people with brown or black skin following these protocols, as diligent use of sunscreen and sun protection in those with darker skin may limit production of vitamin D, potentially causing health problems[19]. Recent scientific studies recognise that typical use of high SPF sunscreens, and sunscreens applied in a thick layer, may have an impact on vitamin D synthesis, and more research has been called for to investigate this[27,28].

What is the SPF?

SPF stands for Sun Protection Factor and indicates the protection against UVB. Product tests to confirm the SPF of a sunscreen involve applying it at 2mg per cm^2 of body surface area, on at least 10 human subjects with Fitzpatrick skin phototype 1 or 2 (as explained in Chapter 1), with a light source that mimics the solar spectrum, to determine the minimal dose of light that produces redness on the individuals' protected and unprotected skin (minimal erythema dose, MED)[11].

The SPF is then calculated as: $$\frac{\text{MED at protected}}{\text{MED at unprotected skin}}$$

If properly applied and tested, a product with SPF 10 should allow 10 times longer in the sun without burning than unprotected skin[11].

What is the UVA Star System?

Developing a test for UVA protection was more difficult than for UVB, as skin redness is not a good indicator of UVA skin damage[11]. Instead, protection can be judged by measuring the overall tanning response, or by dissolving the sunscreen in a solvent and calculating the absorption spectrum in a machine called a spectrophotometer[11].

The UVA rating is shown as a star guide ranging from 0-5 and indicates the percentage of UVA absorbed by the sunscreen in comparison to the UVB. A high number of stars indicate a good balance between UVB and UVA protection, but it is important to note that a low SPF product may still have a high number of stars, as the ratio between the UVA and UVB protection is about the same, but the overall protection is lower[11].

Sunbeds

Commercial sunbeds became common in the late 20th century. In America in 1981, roughly 10 new tanning centres opened each week[9]. Their main purpose is to give you a tan, but they have also been advertised as a source of vitamin D. Sunbeds emit more UVA than UVB as it was initially thought that UVB was more carcinogenic than UVA[29,30]. Unfortunately, UVA is likely to be more carcinogenic than was thought. A person's risk of skin cancer increases with the number of years of sunbed use, the number of sessions and total time of exposure. The lifetime risk of developing melanoma may increase considerably in those people who use sunbeds before the age of 35[31].

How often a sunbed is used and how long for is almost entirely up to the paying individual and some people will put themselves at far more risk than others. Data from a small study of 23 Norwegians in winter indicated that small, non-burning doses of a UVA sunbed twice per week for 15 sessions produced a 25-30% increase in vitamin D[29]. But this should be weighed against the potential cancer risk depending on your skin type. While sunbed use may increase vitamin D, many light skinned people gain enough vitamin D for a healthy body throughout the year by casual sun exposure and taking supplements, and there is little to no obvious support in salons to help individuals safely judge how much UV light is sufficient[18,24].

Fake, Don't Bake

If you do enjoy having a tan, consider using fake tan. Dihydroxyacetone (DHA) is the active chemical most commonly used in fake tan[11]. It is colourless, but binds to the stratum corneum and reacts with a protein to form brown-black compounds called melanodins[32]. The Scientific Committee on Consumer Safety, part of the European Commission, advise that DHA-containing fake tan products are safe and do not pose a risk to health[32]. Fake tan provides a very low SPF and offers some protection from longer wavelengths of radiation into the range of visible light[11].

Vitamin D

Vitamin D can be synthesised from UVB light all year round at southern European and equatorial latitudes, but at the northern latitudes of the UK most of our vitamin D is generated from exposure to UVB light, when it is available between April and September. UVB levels in June here are up to 100 times higher than they are in December[4,5,6,33]. The highest doses of UVB occur in the middle of the day, between 11am and 3pm[4,5,7,24,34,35]. We do get a small amount of vitamin D3 and Vitamin D2 in our diet, and both vitamin forms can be ingested as supplements, but this does not contribute significantly to our levels[36]. As a result, while vitamin D has a half-life of around 4-6 weeks, blood levels can become low by the end of a long northern winter[33].

❖

At northern latitudes most of our vitamin D is generated from exposure to UVB between April and September[4,5,6]

❖

Because our vitamin D level is largely determined by how much UVB penetrates the top layer of our epidermis, it will vary depending on where we live, our personal characteristics like skin type and age, and the way we live our lives, in what we wear and how often we go outside. The results from the National Diet and Nutrition Survey up to 2012 suggested that the risk of vitamin D deficiency has increased across the UK population and almost one fifth of UK adults may have a low vitamin D status[37,38].

People at risk of vitamin D deficiency at northern latitudes include premature babies, children younger than 5 years, pregnant women, those of us with darker skin pigmentation and those older than 65 years[4,38]. The National Institute of Clinical Excellence (NICE) has recommended that health professionals should advise supplement use among at-risk groups whenever possible.

Vitamin D and Health

As discussed in Chapter 1, vitamins D2 and D3 are metabolised in the liver and kidney to form calcitriol, the active form of vitamin D. Calcitriol is a steroid hormone that binds to vitamin D receptors in cell nuclei[39]. There is a receptor in nearly every tissue and thousands throughout the genome controlling hundreds of genes, which gives the potential for vitamin D to be involved in many disease processes[39].

> Almost one fifth of UK adults may have a low vitamin D status[38]

Vitamin D deficiency can result in rickets or osteomalacia, but very low vitamin D levels can also be present without any signs or symptoms[4].

While the links between vitamin D deficiency, osteomalacia and rickets are well established, knowledge and advice on the role of vitamin D in other aspects of health is evolving. The scientific literature in this area is extensive, sometimes contradictory and prone to potentially muddling, confounding factors.

It isn't certain exactly how much vitamin D is needed in the body to stay healthy. The cut-off values indicating vitamin D deficiency and sufficiency vary internationally and between expert groups, and levels can be measured in either nmol/L or ng mL^{-1}. The Institute of Medicine (IOM) guidelines are that levels less than or equal to 30nmol/l (12ng mL^{-1}) are deficient; levels between 30 and 50nmol/L are insufficient; and levels greater than or equal to 50nmol/L (20ng mL^{-1}) are sufficient[40]. The risk of rickets or osteomalacia is higher at levels below 25nmol/L (10ng mL^{-1})[5].

When low vitamin D levels are found in association with a disease it is important to determine whether the vitamin D level is the true cause of the disease, or just related to the circumstances created by it. An example would be, a person with a long history of heart or lung disease, who developed vitamin D deficiency from being unable to go outside into the sunshine. Vitamin D deficiency is occurring in association with the disease, but is not the cause of it, and supplementing the person's vitamin D intake is not likely to resolve their original medical problem.

Within the skin, vitamin D deficiency has been found to be associated with psoriasis, and topical vitamin D derivatives and UV light are used to treat psoriasis[1]. Vitamin D may also be converted in the skin to its active form, calcitriol, which may have a role in improving DNA repair and lowering the risk of skin cancer[1].

A study published in 2014 reviewed current, comprehensive scientific literature (meta analyses and systematic reviews) on the associations between vitamin D levels and 137 medical conditions and diseases. The conclusions indicated probable associations between vitamin D supplementation and higher infant birth weight; a reduction in dental caries in children, and higher vitamin D levels in pregnant mothers at term. There was suggestive evidence for a link between high vitamin D status and lower risk of colorectal cancer, some fractures, cardiovascular disease, high blood pressure, stroke, cognitive function, depression, high BMI, metabolic syndrome, type 2 diabetes and gestational diabetes, and small infant birthweight[41].

But, the authors felt that evidence from more studies and better designed trials was needed for more certainty, and that there was a gap in the quality of evidence concerning vitamin D status and autoimmune diseases, cancer, cognitive function and infectious diseases. Overall, the authors did not feel that the evidence they had seen could support any recommendations about daily vitamin D supplementation for the benefit of general health[41].

Vitamin D and UVB

The dose of vitamin D-creating UVB that hits our skin depends on the height of the sun (solar zenith angle) and hours of sunshine - which vary with season, time of day and latitude - and also the weather, presence of reflective surfaces, pollution and altitude[7,24,36]. The UK does not have a climate that guarantees sunshine everyday, but because vitamin D stays in the body for a few weeks, it is not necessary to receive a UVB dose everyday[7,33]

The amount of UVB available varies with even small changes in latitude[6,33]. Scotland lies within the latitudes 54.3°North to 60.5°North and even within that range there are significant regional differences in UVB dose throughout the year, with the South generally receiving more UVB each month than the North[33]. In line with this, a study testing vitamin D levels of light skinned (Fitzpatrick phototype 1–2) women in the spring found that more women in Aberdeen were vitamin D deficient (21%) than those tested in the south of England (8%) at a latitude of roughly 51.3°North[18].

❖

Our dose of UVB determines how much vitamin D we make … the South of the UK generally receives more UVB than the North[33]

❖

Because of the dependence on UVB, most of us have seasonal variation in our vitamin D levels, but as long as our levels stay above deficiency it is not clear whether this has an impact on our health, or what the impact is. Given the nature of evolution and the length of time that our race has lived at northern latitudes, it could be that our bodies expect some seasonal variation.

NICE recommend that when possible, only a limited amount of time should be spent in strong sunlight to reduce the risk of sun damage and skin cancer[24]. Casual exposure to UK summer sunlight provides enough vitamin D for most, but not all, light skinned 4-64 year olds[18,35]. One UK research study found that within a light skinned study population, being outdoors everyday at lunchtime for 9 (+/- 13) minutes over the summer, with double that time at weekends, resulted in 75% of participants being sufficient in vitamin D between June and October, and by January only 40% were sufficient[35]. Time in the sun outwith midday hours may be increased by walking to and from work for example, and there is less risk of sunburn when out in the sun at this time, but it also takes longer to synthesise vitamin D[24].

❖

Darker skin needs to be exposed to UVB for longer in order to make enough vitamin D[24,34,38]

❖

Darker skin needs to be exposed to UVB for a longer time in order to make enough vitamin D, and those of us with brown or black skin are likely to need a higher summer level of vitamin D in order to remain sufficient through the winter[24,34,38,42]. In Boston (42.2°North) study data indicated that the time needed to obtain enough UVB for adequate vitamin D synthesis each day was 16 minutes for Fitzpatrick skin phototype 1, 49 minutes for phototypes 4-5, and

83 minutes for phototype 6[34]. Times in the UK may be even longer given our higher latitude.

Vitamin D, our Characteristics & Lifestyle

Results from research studies into vitamin D levels can be expected to vary depending on the demographics and characteristics of the study population. Populations of white skinned, outdoorsy young adults are more likely to have sufficient vitamin D levels from short episodes of sun exposure than populations of darker skinned women, children or the elderly. Those of us with brown or black skin, those at the extremes of age, or who lead an overly indoors lifestyle tend to have lower vitamin D levels[18,38].

❖

Almost 75% of Asian adults in the UK may have a low vitamin D status in the winter[38]

❖

Skin Colour

While sunburn can affect any skin type, sun damage and skin cancer from UV rays are largely a problem for people with lighter skin. At northern latitudes, those with genetically darker skin are at a lower risk of sunburn and skin cancer but a substantially higher risk of vitamin D deficiency[19,24,42]. Study data indicates that peak summer vitamin D levels in populations of South Asian ethnicity with brown to dark brown skin living in the UK may be on average 38% of the levels measured in light skinned participants. But, with more time in the sun, darker skin can form just as much vitamin D as lighter skin can[18,42].

Almost 75% of Asian adults with brown to dark brown skin in the UK may have a low vitamin D status in the winter. People with black skin could be expected to have similar or even higher rates[38]. In 2013 a study from Manchester (53.5°North) on a South Asian population found that at no time of the year did more than 7% of the study population reach vitamin D sufficiency. During the winter months 90% had levels below 10ng mL^{-1}, which is low enough to be at risk of osteomalacia if the level is sustained[42].

❖

Older age is associated with lower vitamin D levels

❖

In spring in the south of England, a study comparing the vitamin D levels of South Asian women to that of women from northern Europe with light skin found that 76% of the South Asian women had a vitamin D level that put them at risk of osteomalacia, compared to 8% of women with light skin living in the same area and 21% of light skinned women living in Aberdeen[18]. Problems may not develop if levels rise again through the summer, but, while most light skinned populations across the UK show seasonal variation in vitamin D levels, levels in South Asian women were found to vary very little[18].

Age

Older age is associated with lower vitamin D levels because of a combination of lifestyle and metabolic factors[4,38,43,44]. Some of the highest rates of deficiency can be seen in people older than 65 years[37]. People that are less able to leave their homes will receive less UVB and therefore generate less vitamin D, and vitamin D production in the skin also reduces with age[38,43]. In a study of over 5000 light skinned people aged 60 and above who were tested throughout the year, 32% were deficient or at high risk of deficiency[44]. Of those not taking supplements, people who avoided the sun were typically more at risk of deficiency, while those who enjoyed the sun tended to be sufficient[44].

Lifestyle and Clothing

Many of us have an indoor lifestyle even during summer because of our work, transport habits, electronic entertainment and indoor gym memberships. Unfortunately light filtered through window glass contains very little UVB[4]. Being able to go outside has an obvious positive effect on vitamin D levels, but those of us that are unable to go outside for health, social, psychological or enforced reasons are at risk of vitamin D deficiency.

> More vitamin D can be created within a given time by exposing a larger surface area of skin[7,13]

Outdoor workers, students and young children are the populations best able to be outside in the daytime. In order to keep children safe in the sun many schools and nurseries take part in sun protection protocols involving use of high factor sunscreen, sun hats and shaded play between April and September. As earlier, while sunscreen protects skin from the harmful effects of sunlight, sunscreens may affect vitamin D production when applied in a thick layer and/or at high SPF (30 to 50), especially when the UV index is low but sun protection is advised[27,28]. While those with lighter skin are more at risk of sunburn and still likely to generate vitamin D despite these measures, those with darker skin may be adversely affected by these protocols.

Exposing a larger surface area of skin to the sun will allow it to create a larger amount of vitamin D within a given time, with a lower risk of sunburn[7,13]. The climate in the north of England and Scotland however means that people tend to cover up for much of the year and little skin is regularly exposed to sunlight. Many of us also cover our skin year-round for religious or cultural preference. But, particularly in at-risk groups, wearing clothing like shorts, skirts and short sleeves when possible will encourage more production of vitamin D.

The desire to sit out in the sun is also very relevant for vitamin D levels. Studies have found that, despite a similar time being spent outside, genetically lighter skinned people from northern latitudes tend to seek out the sun and avoid the shade more than those with darker skin, who are of course more in need of sunshine for vitamin D[4,42].

❖
Genetically lighter skinned people seek out the sun more than those with darker skin[42]
❖

Unsurprisingly, having an annual sunny holiday can boost vitamin D levels. In a study population in the north of Scotland, 20-25% of women tested were deficient in vitamin D in winter and spring and 10% were also in summer, but among those who reported a holiday abroad none were deficient in the summer and less than 7% were for the rest of the year[18]. Danish and Swedish studies have also associated sunny holidays with higher vitamin D levels, with evidence that taking a sunny holiday in winter can increase vitamin D levels by almost 15nmol/l[25,26].

Taking part in exercise is associated with higher vitamin D levels, but this result may be confounded by sun exposure as exercise may take place outdoors[4,25]. A study of almost 5000 31 year olds from Northern Finland (65°North), before food fortification was introduced in 2002, found that while the majority were sufficient in vitamin D, the risk of lower vitamin D status was higher in those who had larger waist circumference, who weighed more, followed an unhealthy diet and had more leisure time computer use[45].

Diet and Vitamin D Supplements

While UVB is our main source of vitamin D, obtaining vitamin D through our diet is important when sun exposure is low[4]. Unfortunately, the typical UK diet does not contain much vitamin D, not more than about 3-4mcg (131IU) each day, far less than the estimated ideal daily intake and our intake does not vary through the year[35]. Because of this, once UVB exposure reduces in autumn there is a continuing decline in our vitamin D levels[35].

> ❖
> The typical UK diet does not contain much vitamin D, and our intake does not vary throughout the year
> ❖

Research indicates that despite lower availability of UVB, northern European populations at latitudes higher than 60°North may have better vitamin D status than those at 45 to 60°North because of differences in diet and lifestyle[6]. Eating foods rich in vitamin D can offset limited UVB, for example adults and young people living in Tromso, Norway (69.6°North) have lower levels of vitamin D deficiency and less seasonal fluctuation than estimates from Ireland, which is attributed to the Norwegian intake of fatty fish, cod liver oil and other supplement sources[4,6]. One survey of adults in Norway in 2011 indicated mean vitamin D intakes of around 9-12mcg per day from all sources, with around 7mcg from food alone, three times as much as in other western European countries[4,6]. One herring alone contains 700-800IU (17.5-20mcg) of vitamin D[4]. A Swedish study has suggested that eating fatty fish 2-3 times a week may increase vitamin D levels by up to 45%[26].

In the UK we tend to eat less of these types of fish and very few foods are fortified with vitamin D, so any extra dietary intake really depends on whether we take supplements. The advice on vitamin D supplements in the UK has changed over the years for several reasons: there has been scientific disagreement on what the 'normal' range of vitamin D is; what the recommended daily intake of the vitamin should be; whether vitamin D is important for health apart from bone mineralisation; and variations in the proportional risks of vitamin D deficiency vs. skin cancer between population groups[36].

❖
Eating fatty fish 2-3 times a week may increase vitamin D levels by up to 45%[26]
❖

The UK Scientific Advisory Committee on Nutrition (SACN) updated their advice on vitamin D supplementation in 2016 in the context of public health advice to stay out of the sun in the middle of the day and wear sunscreen[5]. In light of this, the UK government (through the NHS.UK website) and the Chief Medical Officer for the Scottish Government adjusted their recommendations, as described in the table below[46,47].

The biggest change from previous advice was that babies aged between birth and 1 year, drinking less than 500ml formula per day, should also have a daily supplement. Breast milk and breastfeeding provide considerable advantages, including conveying immunity to the baby and reducing the risk of atopic eczema, but breast milk does not contain much vitamin D and the vitamin D it does contain is dependent on the mother's levels[47,48].

UK Advice on Vitamin D Supplementation[46,47]

Population	Advice
Those at risk of deficiency: • Pregnant and breastfeeding mothers • Those not exposed to regular sunlight • Those from ethnic groups with dark skin • Children aged 1-4 years •	10mcg (400IU) daily all year round
The general population above the age of 5 years	Consider taking 10mcg daily particularly over the autumn and winter months
Babies from birth to 1 year	8.5-10mcg daily, unless consuming more than 500ml formula per day

An Irish study suggests that taking any dose of vitamin D supplement can increase levels by 35-40nmol/L but a Scottish study reported that just 26% of its almost 2000 participants reported actually taking supplements[33,44]. To compound the problems with vitamin D deficiency in South Asians and likely other darker skinned populations, vitamin D in the South Asian diet may be just 30-50% of that found in the western pattern diet and only 7% of a South Asian study population recorded having ever taken vitamin D supplements[42].

It is possible to develop vitamin D toxicity from excessive dietary supplementation. The maximum levels for adults and children over the age of 11 have been reported as 100mcg per day, children from 1 to 10 a maximum of 50mcg per day and infants below 1 year no more than 25mcg per day[46].

Vitamin D, Pregnancy and Infants

Pregnancy

A pregnant woman's vitamin D level determines the level in the newborn baby[47,49,50]. The incidence of vitamin D deficiency in pregnancy varies between 18% and 84% of expectant mothers, depending on area of the world and clothing customs[49]. Low maternal vitamin D levels have been associated with preeclampsia, small infant birthweight and health problems in infants[49,50].

The current UK recommendation is that all pregnant and nursing mothers should take a vitamin D supplement of 400IU (10mcg daily), regardless of time of year[46]. Multivitamin preparations contain vitamin D, but many preparations for general consumption are not suitable for pregnant women because of their vitamin A content. The vitamin preparations designed for use in pregnancy contain vitamin D as well as safe levels other supplements.

> Pregnant women who have migrated from Africa and Asia to the Northern Hemisphere often have very low vitamin D levels[4,50]

Pregnant women who have migrated from Africa and Asia to northern latitudes often have very low vitamin D levels[4,50]. A Swedish study investigated vitamin D levels in over 2000 pregnant women. 75% of the group were born in Northern Europe, 6% in Africa and 10% in Asia[50]. While overall only 10% of the study group had deficient Vitamin D levels, deficiency was common in women with darker skin – found in 51% of those born in Africa and 46% of those born in Asia[50].

Infants

A newborn baby's vitamin D level is determined by the mother's level during pregnancy[47,49,50]. As above, breast milk only contains small amounts of vitamin D and these levels are even lower in deficient mothers[38,51,52]. Formula milks are fortified with vitamin D, but infants who have less than 500ml formula per day may not get enough vitamin D for their needs[5,38,46,47,52]. At the same time, public health advice recommends that children under 6 months should be kept out of direct strong sunlight, and between March and October children and young people should have their skin protected with suitable clothing, be encouraged to spend time in the shade and wear sunscreen[24]. The updated advice on supplement intake goes some way to combatting this issue.

❖

The number of children suffering from rickets has been rising across the world[38,53]

❖

Vitamin D deficiency is quite common in children, with Afro-Caribbean and Asian groups at particular risk, as well as those children with problems absorbing vitamin D across the gut, or those less able to go outside[51]. The rate of vitamin D deficiency in newborns with brown to black skin has been reported as up to 46%, compared with just 10% in newborns with light skin, and the Asian Feeding Survey has reported that up to one third of two year old Indian, Bangladeshi or Pakistani children may have low vitamin D status[38,49].

Children can be deficient in vitamin D without displaying any symptoms, but there is concern that the number of children suffering from rickets is rising across the world[38,53]. Rickets may present with weakness of the proximal muscles, especially the thigh muscles (known as Gower's sign) and difficulty walking, swollen wrists and eventually the classic feature of bowed legs. Teenagers may present with aches and pains in their legs[51].

Sunshine and Vitamin D: Balancing Risks and Benefits

With its sunny climate and large fair skinned population, Australia has lead the way in public health advice on sun safety. Australia has one of the highest rates of skin cancer in the world. Two out of three Australians will develop a form of skin cancer before age 70 and sun exposure in childhood has a significant impact on a person's risk[54].

Public health and educational campaigns launched in Australia in the 1980's significantly altered attitudes and behaviours towards sun exposure and the attractiveness of the suntan[9]. This change in attitude was very important as the country has high levels of UV, many sunny days and a culture of outdoor activity. But northern climates are very different to Australia and stepping outside on an average spring day in Glasgow will confirm that, sadly, you are not in Australia.

The results of vitamin D research studies are affected by the study location and latitude as well as the characteristics of the study population[7]. Results can vary considerably depending on these variables, which may cause confusion when translating results into public advice. For example, there is evidence that sufficient vitamin D levels can be produced by just a few minutes of sun exposure to only the face and hands, but this and similar studies were performed at low or equatorial latitudes, under local sun intensity and weather conditions[7]. Their conclusions would not be appropriate for use in planning health advice for northerly populations.

While emphasising the very real risks of excessive sun exposure and skin cancer to the public, UK health advice has in the past been similar to that given in far sunnier countries with little mention of the impact that differences in skin type or circumstances can have[42]. Scientific papers had called for adjustments to be made to public health advice, with more consideration and guidance on how all of the population could maintain sufficient vitamin D levels while remaining safe in the sun[35].

More recently, in 2016 NICE recommended that public messages about the risks and benefits of sun exposure should be consistent and balanced. They advised that campaigns should address health inequalities by taking into account the effect of cultural, religious and social behaviours on sun exposure, as well as outlining what different population groups can do to reduce their risk - whether that risk is sun damage or vitamin D deficiency[24].

Public health policies and sun protection protocols provide advice aiming to keep as many of us as possible safe and healthy. It would be unrealistic to ask a school or nursery to apply their sun protection policies differently for each child in their care depending on their circumstances. But in our own time as individuals we can all be aware of our own skin type and that of people in our care, and take reasonable measures to remain sufficient in vitamin D while staying safe in the sun.

Summary

- Sunshine, skin and health have a complicated relationship that centres on balance.
- Occasional exposure to strong sunlight carries an increased risk of sunburn, with a knock-on increased risk of skin cancer.
- In the UK, most of our vitamin D is generated from exposure to UVB light in the middle of the day, between April and September.
- Vitamin D supplements should be encouraged for those people at risk of deficiency.
- The balance between risk of sun damage and skin cancer, and risk of vitamin D deficiency varies with our age, skin type and way of life.
- Regular non-burning sun exposure carries health benefits independent of vitamin D.
- Public messages about the risks and benefits of sun exposure should be consistent, balanced and inclusive.

References

1. Hoel DG, Berwick M, de Gruijl FR, Holick MF. The risks and benefits of sun exposure 2016. Dermatoendocrinol. 2016;8(1):e1248325
2. Weller RB. Sunlight has cardiovascular benefits independently of vitamin D. Blood Purif. 2016;41(1-3):130-134
3. Villanueva JC. Radiation from the Sun [Internet]. Canada: Universe Today, space and astronomy news; [2010 Mar.; cited 2019 June]. Available from: https://www.universetoday.com/60065/radiation-from-the-sun
4. Lips P, van Schoor NM, de Jongh RT. Diet, sun and lifestyle as determinants of vitamin D status. Ann N Y Acad Sci. 2014;1317:92-98
5. Scientific Advisory Committee on Nutrition (SACN). Vitamin D and Health. 2016 July. Available online from: https://assets.publishing.service.gov.uk/government/uploads/system/uploads/attachment_data/file/537616/SACN_vitamin_and_health_report.pdf
6. O'Neill CM, Kazantzidis A, Ryan MJ, Barber N, Sempos CT, Durazo-Arvizu RA, Jorde R, Grimnes G, Eiriksdottir G, Gudnason V, Cotch MF, Kiely M, Webb AR, Cashman KD.Seasonal changes in vitamin D-effective UVB availability in Europe and associations with population serum 25-hydroxyvitamin D. Nutrients. 2016;8(9). Pii: E533 doi: 10.3390/nu8090533
7. McKenzie RL, Liley JB, Bjorn LO. UV radiation: balancing risks and benefits. Photochem Photobiol. 2009;85(1):88-98
8. Zarrelli N. The poisonous beauty advice columns of Victorian England [Internet]. United States of America: AltasObscura.com; [2015 Dec.; cited 2019 June]. Available from: https://www.atlasobscura.com/articles/the-poisonous-beauty-advice-columns-of-victorian-england
9. Chang C, Murzaku EC, Penn L, Abbasi NR, Davis PD, Berwick M, Polsky D . More skin, more sun, more tan, more melanoma. Am J Public Health. 2014;104(11):e92-99
10. Hockberger PE. A history of ultraviolet photobiology for humans, animals and microorganisms. Photochem photobiol. 2002;76(6):561-579
11. Bolognia JL, Jorizzo JL, Schaffer JV. Dermatology. 3rd ed. United States of America: Elsevier Saunders; 2012

12. Albert MR Ostheimer KG. The evolution of current medical and popular attitudes toward ultraviolet light exposure: part 1. J Am Acad Dermatol. 2002;47(6):930-937
13. Miyauchi M, Nakajima H. Determining an effective UV radiation exposure time for vitamin D synthesis in the skin without risk to health: simplified estimations from UV observations. Photochem Photobiol. 2016;92(6):863-869
14. Addor FA. Antioxidants in dermatology. An Bras Dermatol. 2017;92(3):356-362
15. The Met Office. UV Index Forecast [Internet]. United Kingdom: The Met Office; [unknown; cited 2019 Jun.]. Available from: https://www.metoffice.gov.uk/public/weather/uv-indexforecast/#?tab=map&map=MaxUVIndex&zoom=6&lon=-4.00&lat=52.83&fcTime=1560416400
16. Holloway L. Shadow method for sun protection. Lancet. 1990;335(8687):484
17. Gandini S, Sera F, Cattaruzza MS, Pasquini P, Picconi O, Boyle P, Melchi CF. Meta-analysis of risk factors for cutaneous melanoma: II. Sun exposure. Eur J Cancer 2005;41(1):45-60
18. Ashwell M, Stone EM, Stolte H, Cashman KD, Macdonald H, Lanham-New S, Hiom S, Webb A, Fraser D. UK Food Standards Agency workshop report: an investigation of the relative contributions of diet and sunlight to vitamin D status. Br J Nutr. 2010;104(4):603-611
19. Diffey BL, Fajuyigbe D, Wright C. Sunburn and sun protection in black skin. Int J Dermatol. 2019;doiL10.1111/ijd.14402 [Epub ahead of print]
20. Lazovich D, Vogel RI, Berwick M, Weinstock MA, Anderson KE, Warshaw EM. Indoor tanning and risk of melanoma: a case control study in a highly exposed population. Cancer Epidemiol Biomarkers Prev. 2010;19(6):1557-1568
21. Matta MK, Zusterzeel R, Pilli NR, Patel V, Volpe DA, Florian J, Oh L, Bashaw E, Zineh I, Sanabria C, Kemp S, Godfrey A, Adah S, Coelho S, Wang J, Furlong LA, Ganley C, Michele T, Strauss DG. Effect of sunscreen application under maximal use conditions on plasma concentration of sunscreen active ingredients: a

randomised clinical trial. JAMA. 2019; doi 10.1001/JAMA.2019.5586 [epub ahead of print]

22. Wang J, Ganley CJ. Safety threshold considerations for sunscreen systemic exposure: a simulation study. Clin Pharmacol Ther. 2019;105(1):161-167

23. Ruszkiewicz JA, Pinkas A, Ferrer B, Peres TV, Tsatsakis A, Aschner M. Neurotoxic effect of active ingredients in sunscreen products, a contemporary review. Toxicol Rep. 2017;4:245-259 doi: 10.1016/j.toxrep.2017.05.006

24. National Institute for Health and Care Excellence (NICE). Sunlight exposure: risks and benefits NICE guideline ng34. 2016 Feb [cited 2019 June]. Available online from: https://www.nice.org.uk/guidance/ng34

25. Madsen KH, Rasmussen LB, Mejborn H, Andersen EW, Molgaard C, Nissen J, Tetens I, Andersen R. Vitamin D status and its determinants in children and adults among families in late summer in Denmark. Br J Nutr. 2014;112(5):776-784

26. Burgaz A, Akesson A, Oster A, Michaelsson K, Wolk A. Associations of diet, supplement use, and ultraviolet B radiation exposure with vitamin D status in Swedish women during winter. Am J Clin Nutr. 2007;86(5):1399-1404

27. Neale RE, Khan SR, Lucas RM, Waterhouse M, Whiteman DC, Olsen CM. The effect of sunscreen on vitamin D: a review. Br J Dermatol. 2019;doi: 10.1111/bjd.17980 [epub ahead of print]

28. Young AR, Narbutt J, Harrison GI, Lawrence KP, Bell M, O'Connor C, Olsen P, Grys K, Baczynska KA, Rogowski-Tylman M, Wulf HC, Lesiak A, Philipsen PA. Optimal sunscreen use, during a sun holiday with a very high ultraviolet index, allows vitamin D synthesis without sunburn. Br J Dermatol. 2019; doi: 10.1111/bjd.17888 [epub ahead of print]

29. Moan J, Lagunova Z, Cicarma E, Aksnes L, Dahlback A, Grant WB, Porojnicu AC. Sunbeds as vitamin D sources. Photochem Photobiol. 2009;85(6):1474-1479

30. The European Committee for Electrotechnical Standardization (CENELEC). Household and similar electrical appliances – safety – part 2-27: particular requirements for appliances for skin exposure to ultraviolet and infrared radiation (IEC 60335-2-27:2009 modified). 2013 Dec. Available online from:

http://ec.europa.eu/health/scientific_committees/scheer/docs/sunbeds_co44_en.pdf
31. D'Orazio J, Jarrett S, Amaro-Ortiz A, Scott T. UV radiation and the skin. Int J Mol Sci. 2013;14(6):12222-12248
32. Scientific committee on consumer safety (SCCS). Opinion on dihydroxyacetone SCCS/1347/10. 2010 Dec. Available online from: https://ec.europa.eu/health/scientific_committees/consumer_safety/docs/sccs_o_048.pdf
33. Kelly D, Theodoratou E, Farrington SM, Fraser R, Campbell H, Dunlop MG, Zgaga L. The contributions of adjusted ambient ultraviolet B radiation at place of residence and other determinants to serum 25-hydroxyvitamin D concentrations. Br J Dermatol. 2016;174(5):1068-1078
34. Webb AR, Engelsen O. Calculated ultraviolet exposure levels for a healthy vitamin D status. Photochem Photobiol. 2006;82(6):1697-1703
35. Webb AR, Kift R, Durkin MT, O'Brien SJ, Vail A, Berry JL, Rhodes LE. The role of sunlight exposure in determining the vitamin D status of the UK white adult population. Br J Dermatol. 2010;163(5):1050-1055
36. Laird E, Ward M, McSorley E, Strain JJ, Wallace J. Vitamin D and bone health: potential mechanisms. Nutrients. 2010;2(7):693-724
37. Public Health England, Food Standards Agency. The national diet and nutrition survey: results from years 1,2,3 and 4 (combined) of the rolling programme (2008/2009 – 2011/2012). 2017 Feb. Available online from: https://assets.publishing.service.gov.uk/government/uploads/system/uploads/attachment_data/file/594360/NDNS_Y1_to_4_UK_report_executive_summary_revised_February_2017.pdf
38. National Institute for Health and Care Exellence (NICE). Vitamin D: supplement use in specific population groups NICE guidance ph56. 2014 Nov. [cited 2019 June]. Available online from: http://www.nice.org.uk/guidance/ph56
39. Bikle DD. Vitamin D metabolism, mechanism of action, and clinical applications. Chem Biol. 2014;21(3):319-329
40. Ross AC, Manson JE, Abrams SA, Aloia JF, Brannon PM, Clinton SK, Durazo-Arvizu RA, Gallagher JC, Gallo RL,

Jones G, Kovacs CS, Mayne ST, Rosen CJ, Shapses SA. The 2011 report on dietary reference intakes for calcium and vitamin D from the Institute of Medicine: what clinicians need to know. J Clin Endocrinol Metab. 2011;96(1):53-58

41. Theodoratou E, Tzoulaki I, Zgaga L, Ioannidis JP. Vitamin D and multiple health outcomes: umbrella review of systematic reviews and meta-analyses of observational studies and randomised trials. BMJ. 2014;348:g2035 doi: 10.1136/BMI.g2035

42. Kift R, Berry JL, Vail A, Durkin MT, Rhodes LE, Webb AR. Lifestyle factors including less cutaneous sun exposure contribute to starkly lower vitamin D levels in UK South Asians compared with the white population. Br J Dermatol. 2013;169(6):1272-1278

43. Mukherjee S, Date A, Patravale V, Korting HC, Roeder A, Weindl G. Retinoids in the treatment of skin aging: an overview of clinical efficacy and safety. Clin Interview Aging. 2006;1(4):327-348

44. O'Sullivan F, Laird E, Kelly D, van Geffen J, van Weele M, McNulty H, Hoey L, Healy M, McCarroll K, Cunningham C, Casey M, Ward M, Strain JJ, Molloy AM, Zgaga L. Ambient UVB dose and sun enjoyment are important predictors of vitamin D status in an older population. J Nutr. 2017;147(5):858-868

45. Palaniswamy S, Hypponen E, Williams DM, Jokelainen J, Lowry E, Keinanen-Kiukaanniemi S, Herzig KH, Jarvelin MR, Sebert S. Potential determinants of vitamin D in Finnish adults: a cross-sectional study from the Northern Finland birth cohort 1966. BMJ Open. 2017;7(3):e013161 doi: 10.1136/bmjopen-2016-013161

46. National Health Service (NHS). Vitamin D [Internet]. United Kingdom: NHS; [2017 Mar.; cited 2019 June]. Available from: https://www.nhs.uk/conditions/vitamins-and-minerals/vitamin-d

47. Calderwood C. Correspondence: new recommendations on vitamin D supplementation. Chief Medical Officer Directorate. 2017 Nov. Available online from: https://www.cps.scot/media/1987/cmo-unnumbered-letter-issued-on-24-november-2017-final-new-recomme.pdf

48. Ong KK, Forouhi N. Communicating the benefits of breast feeding. Arch Dis Child. 2007;92(6):471-472

49. Mulligan M, Felton SK, Riek AE, Bernal-Mizrachi C. Implications of vitamin D deficiency in pregnancy and lactation. Am J Obstet Gynaecol. 2010;202(5):429.e1-9 doi:10.1016/j.ajog.2009.09.002
50. Barebring L, Schoenmakers I, Glantz A, Hulthen L, Jagner A, Ellis J, Barebring M, Bullarbo M, Augustin H. Vitamin D status during pregnancy in a multi-ethnic population-representative Swedish cohort. Nutrients. 2016;8(10). Pii: E655
51. Royal College of Paediatrics and Child Health (RCPCH). Position statement: Vitamin D. RCPCH 2012 Dec.
52. Scientific Advisory Committee on Nutrition (SACN). Feeding in the first year of life. SACN. 2018 Jul. Available online from: https://www.gov.uk/government/publications/feeding-in-the-first-year-of-life-sacn-report
53. Uday S, Hogler W. Nutritional rickets and osteomalacia in the twenty-first century: revised concepts, public health, and prevention strategies. Curr Osteoporos Rep. 2017;15(4):293-302
54. Cancer Council Victoria. Be Sunsmart, Play Sunsmart; Early Childhood Resource. Cancer Council Victoria. 2011. Available online from: https://www.sunsmart.com.au/downloads/communities/early-childhood-primary-school/resources/be-sunsmart-play-sunsmart.pdf

6. Jumping Jack Frost:

Skin Problems from the Cold

6. Jumping Jack Frost:
Skin Problems from the Cold

While skin dryness and irritation, rosacea and vitamin D sufficiency might be topics that quickly come to mind when considering cold weather and the skin, there are other skin conditions that might not be particularly common, but are directly related to exposure to cold temperatures. Who knows when we might find it useful to know more about them, and it is particularly helpful for people affected by these conditions if more of us are generally aware of them.

I have separated them roughly as those caused by a cold injury to the skin, including frostbite and cold water immersion foot, and conditions caused by a reaction inside the body, such as cold urticaria, Reynaud's phenomenon and chilblains. Cold panniculitis sits somewhere between these two groups. In this case, inflammation of the subcutaneous fat (known as panniculitis) has happened in direct response to cold, so may be considered an injury, but panniculitis may also develop as part of a skin disease.

Injuries from Cold Exposure

Cold injuries will have occurred for as long as humans have relied on clothing, shelter and fire for warmth. Mass casualties have occurred during wars and times of mass migration. In peacetime day to day life, occupational cold injuries may develop in explorers, military personnel, fishermen and those working in refrigerated areas, as well as in the homeless, those trapped outside during low temperatures, or people taking part in extreme outdoor sports[1,2,3].

Cold injuries will have occurred for as long as humans have relied on clothing, shelter and fire for warmth.

The clothing we wear in the cold obviously affects how comfortable we feel, but also has some surprising effects on our bodies. Physiological studies have found that being out in cold temperatures increases our blood pressure, but wearing a hat reduces the amount that it increases by[3]. While out in the cold, our fingers, nose, cheeks and chin are often the coldest parts of our bodies. Wearing thermal facial protection like a large scarf or balaclava not only warms our face, but can somehow help to keep fingers and hands warm as well and results in us generally feeling more comfortable during exposure to temperatures well below zero[3,4].

It is not surprising that taking part in exercise will help to warm us up when we are cold. Moderate exercise helps us feel more comfortable in a cold wind, with less increase in blood pressure and less cooling at the nose and fingers[3]. The combination of exercise maintaining our core body temperature and cold induced vasodilatation (as described in Chapter 3) keep the regions of the face that stick out, such as the nose, warmer compared to other parts of the face like the forehead or cheek[3,5,6].

Drugs, alcohol, problems with blood vessels, smoking and high altitude all increase the risk of a cold injury[3,6]. Extreme cold may reduce core body temperature, causing hypothermia, or reduce the temperature at peripheral areas like the hands and feet, leading to freezing of the tissue (frostbite) or non-freezing injuries like cold immersion foot[1,3].

Frostbite

Frostbite occurs at air temperatures below 0°C, with freezing of peripheral tissue at the fingers, toes and ears. In cold conditions the risk of frostbite increases with extremes of age, alcohol and drug use and vibration exposure at the fingers. In Finland, depending on latitude, age and occupation, up to 68% of adults may have had frostbite at some point, with the highest numbers among reindeer herders[3].

In very cold and windy conditions it is important to look out for signs of skin cooling, as they indicate impending tissue damage. The earliest stage of frostbite is known as frostnip. When the skin temperature is below 15°C, skin at the area becomes pale and a sharp, stinging pain develops, followed by numbness[1]. Skin may turn pink for a short period but then goes back to white[3].

> ❖
> Frostbite occurs at air temperatures below 0°C, with freezing of peripheral tissues
> ❖

Treatment should start as soon as the affected person is aware they have lost sensation in their skin. Shelter, warm drinks, replacing wet clothing with dry and warming hands or feet in a body flexure like the armpit can help. Avoid rubbing the area or applying direct heat[7]. If possible, rewarm the area in a warm water bath (a whirlpool bath if possible to reduce pain[1]). If the casualty also has hypothermia it is important to warm the core of the body before the limbs to prevent a sudden drop in blood pressure and shock[2].

Degrees of Frostbite[1,2]

First degree frostbite (Frostnip)
There is redness, swelling, numbness and short-lived pain at the area. Only the very top layer of skin is affected and full recovery is expected with just mild skin peeling.

Second degree frostbite
There is significant redness, swelling and blistering. Healing occurs but nerve damage and sensitivity to cold can persist.

Third degree frostbite
There is tissue death affecting the full thickness of the skin, with development of blood filled blisters and dry, waxy tissue.

Fourth degree frostbite
There is tissue death affecting skin, muscle, tendon and bone, leading to amputation.

Active movement during rewarming can be helpful, but frostbitten limbs cannot be used for weight bearing once rewarmed[1,7]. Exercise has been found to be more effective at reheating cooled hands and fingers than putting on extra clothing, and even prior to any injury, the risk of frostbite in the nose is felt to be lower when we are more active[3].

Other treatments include anti-inflammatories like ibuprofen, pain relief, tetanus immunisation, and for severe cases the consideration of hyperbaric oxygen, medications to thin the blood, medications that dilate blood vessels, or surgery to relieve pressure and remove dead tissue[1].

How severe an injury is can be difficult to determine straight away and it can take weeks before the true extent of the damage is seen[1]. If an early injury is treated quickly it can heal with no further issues, but deep cold injuries need a long time to heal and may result in lifelong problems[3]. Avoid further frostbite at the area soon after it is treated, as tissue that is rewarmed but then refrozen almost always dies[7].

Cold Water Immersion Foot

Also known as Trench Foot, this cold injury occurs when tissue is continually immersed and exposed to temperatures between 0-15°C for several hours or days, or when feet are immersed at 15-21°C for more than 7 days[1,2].

The stratum corneum at the sole of the foot can absorb up to 200% of its dry weight in water and it becomes overhydrated, macerated and the skin breaks down[2]. The cold causes blood vessels in the area to spasm, reducing blood flow, resulting in injuries to nerves, muscle, fat and blood vessels themselves because of a lack of oxygen[2]. Feet become cold, stiff and numb and may appear swollen, waxy white to blue in colour.

❖
Anyone who spends long periods of time with wet feet is at risk[1]
❖

Immersion foot has most famously been described in military personnel during wars, but anyone who spends long periods of time with wet feet is at risk[1]. Risk factors include prolonged job or hobby-related exposure to cold and wet conditions, immobility, an altered mental state, dehydration, poor nutrition, smoking, disease of the blood vessels, uncontrolled diabetes and homelessness[1,2].

Immersion foot has been reported in a patient with erythromelalgia, which is a condition causing severe, persisting pain in the lower legs that doesn't respond to many of the usual pain relief medications. People may resort to nerve distraction techniques for relief by applying fans, ice and cold water to the area for long periods of time[8].

Severe injury may produce nerve damage and long-term problems, with risks of infection, muscle wasting, problems with venous circulation and joint contractures[2]. Prevention is the best form of management, but once the condition develops, drying the tissue is essential. Rewarming should be slow to avoid worsening the injury, and amitriptyline seems to work best in terms of pain relief[1].

Cold Panniculitis

Panniculitis means inflammation of the subcutaneous fat. It appears as red to pink, or bruise-like firm swellings under the skin[2,9]. There are several causes of panniculitis, and in some cases it may be related to underlying medical conditions, but in this case it is being caused by cold trauma inflicted on the body.

Cold panniculitis, where inflammation occurs in response to cold exposure, has been described in children as 'popsicle panniculitis' because of its appearance at the cheeks after sucking ice-lollies or exposure to cold[10]. The particular sensitivity children have to this condition may be because of a difference in their ratios of fatty acids compared to those of adults, and cold sensitivity reduces as children get older[10]. It has also been described in horse riders[2], and there has been one case of widespread cold panniculitis described in a 47 year old man who had received whole body cryotherapy. This therapy exposes the body to temperatures below -110°C for several minutes, and is offered in some sectors for improvement in muscle pain after exercise[9]. Cold panniculitis generally gets better within a few weeks with no long-lasting effects[10].

Skin Disease from Cold Exposure

Cold Urticaria

Urticarial rashes appear as itchy, pink to red swellings commonly known as wheals or hives. Urticarial rashes can develop as part of an allergic reaction, because on recognising the allergen the immune system orders mast cells in the dermis to burst, releasing histamine and other inflammatory mediators, producing the rash.

However, many urticarial rashes are not caused by allergy. Mast cells can release their histamine and cause an urticarial rash when no allergen is present. Urticaria can appear after infections, in association with some diseases, by the direct action of some foods and drugs on the mast cells, through inherited conditions, and for unknown reasons (known as idiopathic urticaria).

> Up to around one in three of us will have a form of urticaria at some point in our lives[2]

Up to around one in three of us will have a form of urticaria at some point in our lives and it may develop at any age[2]. Cold urticaria is quite a rare form. In cold urticaria mast cells burst and release histamine in response to the skin coming into contact with cold water, cold objects, cold environments or significant changes in temperature. This type of urticaria can also be inherited, with inherited cases normally first appearing during childhood. Cold urticaria is more common in cold climates, which must be at least in part because it will appear more often[11]. The condition can occasionally progress to anaphylaxis, and people with it are advised not to swim in cold water, to wear appropriate cold weather clothing, and to avoid cold weather activities.

Managing cold urticaria can affect people's working life, stop them from taking part in activities or hobbies and sometimes stop them from leaving their homes, which you can imagine will have an affect on your general happiness and way of life[11]. Antihistamines, mast cell stabilisers and drugs affecting the immune system are used to control and prevent symptoms, and people are likely to be given an Epipen because of the potential risk of anaphylaxis.

Apart from inherited cases, improvement in symptoms or full relief from the condition occurs in 50% of people within 5-6 years, and once the condition has gone people generally remain free of it. In inherited cases however it can last a lifetime[11].

Cryogelling and Cryoagglutination

Cryoglobulins are proteins in the blood that can precipitate on exposure to cold, and cold agglutinins are antibodies that promote clumping of red blood cells, also on exposure to cold. Both of these can cause problems in the skin, or elsewhere in the body, from blockage or inflammation of small blood vessels.

These proteins usually appear in association with other illness, including infections and diseases affecting the immune system. The effect on the blood vessels can produce a lacy or spotty, red / purple rash, which does not disappear on pressure (does not blanch) at the sites of cold exposure. Sometimes skin in the area can blister or break down (ulcerate) because the reduction in blood flow means less oxygen and nutrients are brought to the area.

Raynauds Phenomenon

Reynaud's phenomenon is a change in skin colour, usually at the fingers, caused by a brief spasm of small arteries. It is often triggered by the cold. It is normal for skin to change colour in response to cold, but in Reynaud's the colour shifts noticeably between white, blue and red. Typically, one or more fingers will turn white and numb, then blue on re-warming and then bright red, sometimes with swelling. Attacks can be painful and last from minutes to hours[12].

❖
Nicotine tightens blood vessels[13]
❖

Reynaud's most commonly appears in young women and is not usually associated with other medical problems but can sometimes be triggered by stress. Occasionally, Reynaud's can develop alongside other medical problems such as connective tissue or immune system disease. In this situation there are usually many attacks each day and cold may be the only trigger[2]. Reynaud's may develop in construction workers who have used vibrating tools and also in industrial workers exposed to vinyl chloride polymerisation processes[12].

Symptoms can be reduced by stopping smoking (nicotine tightens blood vessels[13]), heating the home adequately, avoiding activities that trigger attacks, wearing warm clothes including gloves and socks and using hand and foot warmers. Sometimes, medical treatment with nifedipine, diltiazem or other agents is used to try and prevent the vessel spasms. Depending on any associated conditions, Reynaud's may persist for many years or may get better by itself [12].

Chilblains

Chilblains are also known as pernio. They appear as itchy or burning, purplish skin lesions, sometimes with blistering and ulceration. They usually affect the fingers, toes, face, nose and ears and are often symmetrical. Women, children and the elderly are most often affected. They are usually caused by cold, wet conditions and are common in the UK and Northern Europe, particularly in homes with no central heating[2].

The exact processes leading to chilblains appearing are not well known but they are thought to originate from a problem with small blood vessels, with possible leakage of blood cells into the surrounding tissue contributing to inflammation[2,13]. The symptoms and diagnostic criteria for chilblains overlap with the definitions that some organisations hold for mild frostbite[1].

Chilblains can be uncomfortable but with avoidance of the cold they normally heal within a few weeks leaving no permanent damage[13]. The chance of chilblains developing may be reduced by wearing appropriate clothing, avoiding cold and damp conditions, keeping feet and fingers dry and moisturised, warming shoes and gloves, avoiding smoking, keeping active, eating at least one hot meal per day and keeping the house well heated[13]. If this fails, topical steroid creams and nifedipine 20-60mg daily have been used to prevent lesions appearing[2].

Summary

- Cold injuries may often be associated with times of war or mass migration but many people are also at risk in their day-to-day lives, for occupational, recreational and social reasons.
- Prevention is the best form of management.
- Drug and alcohol intake, problems with blood vessels, smoking and high altitude are all risk factors for a cold injury.
- Exercise is more effective at reheating cold hands and fingers than putting on extra clothing.
- Symptoms from skin reactions to the cold, like Reynaud's and chilblains, can be reduced by keeping skin warm and dry, wearing protective clothing, heating your home, eating warm food and avoiding smoking.

References

1. Heil K, Thomas R, Robertson G, Porter A, Milner R, Wood A. Freezing and non-freezing cold weather injuries: systematic review. Br Med Bull.2016;117(1):79-93
2. Bolognia JL, Jorizzo JL, Schaffer JV. Dermatology. 3rd ed. United States of America: Elsevier Saunders; 2012
3. Gavhed D. Human responses to cold and wind. PhD Thesis. Department of Physiology and Pharmacology, Karolinska Institutet, Stockholm, Sweden. 2003

4 O'Brien C, Castellani JW, Sawka MN. Thermal face protection delays finger cooling and improves thermal comfort during cold air exposure. Eur J Appl Physiol. 2011;111(12):3097-3105

5 Brajkovic D, Ducharme MB. Facial cold-induced vasodilatation and skin temperature during exposure to cold wind. Eur J Appl Physiol. 2006;96(6):711-721

6 Daanen HA. Finger cold-induced vasodilatation: a review. Eur J Appl Physiol. 2003;89(5):411-426

7 Imray C, Grieve A, Handford C, Cooper B, Hudson S. Frostbite [Internet]. United Kingdom: theadventuremedic.com; [2013 Jun.; cited 2019 Jun.]. Available from: https://www.theadventuremedic.com/features/frostbite/

8 Davis MD. Immersion foot associated with the overuse of ice, cold water and fans: a distinctive clinical presentation complicating the syndrome of erythromelalgia. J Am Acad Dermatol. 2013;69(1):169-171

9 Greenwald E, Christman M, Penn L, Brinster N, Liebman TN. Cold panniculitis: adverse cutaneous effect of whole-body cryotherapy. J Am Acad Dermatol. 2018;4(4):344-345

10 Huang FW, Berk DR, Bayliss SJ. Popsicle panniculitis in a 5-month old child on systemic prednisolone therapy. Pediatr Dermatol. 2008;25(4):502-503

11 Singleton R Halverstam CP. Diagnosis and management of cold urticaria. Cutis. 2016;97(1):59-62

12 Maurice P. Reynaud phenomenon [Internet]. New Zealand: DermnetNZ.org; [2004; cited 2019 Jun.]. Available from: www.dermnetnz.org/topics/raynaud-phenomenon/

13 National Health Service (NHS). Chilblains [Internet]. United Kingdom: NHS; [2018 Nov.; cited 2019 Jun.]. Available from: www.nhs.uk/conditions/chilblains

7. Keeping it Cool: Kids in the Cold

7. Keeping it Cool:
Kids in the Cold

In children with generally healthy skin and an intact skin barrier, most issues seen in cold weather are down to one or a combination of snot, saliva and generally being a small person with strong feelings. In kids who do have a skin condition, this troublesome trio can create an extra layer of miserable. Maintaining the integrity of the skin barrier is of the greatest importance in protecting children's skin against irritants and cold weather.

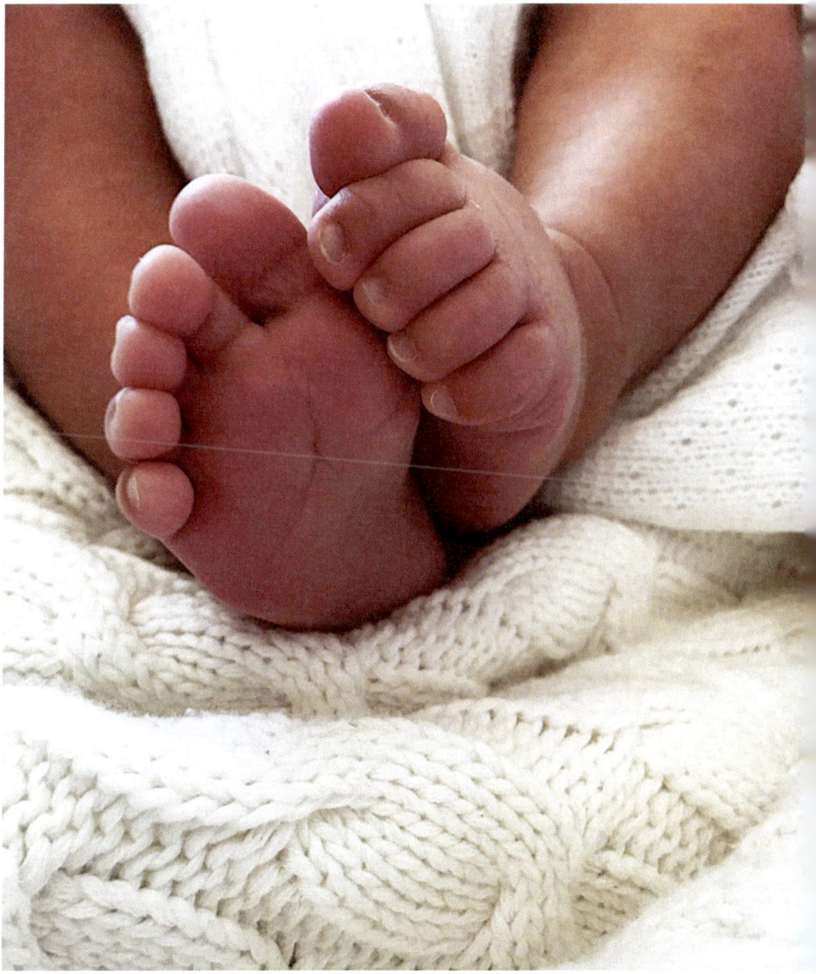

The Skin Barrier

The epidermal skin barrier keeps skin hydrated and prevents entry of bugs, irritants and allergens[1]. Skin structure in children is essentially the same as in adults, but babies under the age of one have a thin, incomplete skin barrier and are more prone to dryness and skin irritation[2,3].

Newborns have a neutral skin surface pH initially, which decreases to become more acidic during the first month of life as the acid mantle forms. The enzymes involved in forming the skin barrier need an acidic environment in order to do their job and so newborn skin is relatively dry while they are waiting for the pH to drop. On the other hand, the enzymes involved in desquamation enjoy a higher pH and so increase their activity, contributing to skin peeling and flaking[4]. After the acid mantle forms and enzyme activity normalizes, in the absence of skin disease, young children have better hydrated skin than adults do[3].

❖
Eczema affects about 20% of children and the integrity of the skin barrier is central to the disease process
❖

Eczema, also known as atopic dermatitis, affects about 20% of children (compared to 5% of adults) and is the most common chronic skin disease in the world[1]. It is strongly linked with the development of food allergy, asthma and hay fever and may occur at the same time as these other atopic conditions[1].

There are many genetic and environmental factors that can trigger or contribute to the development of eczema, but the integrity of the skin barrier is central to the disease process. Cytokine messenger molecules in eczematous skin alter the amounts of proteins and lipids present, leading to the skin barrier becoming more leaky, and more water being lost through the epidermis. Irritants and allergens can enter the skin more easily and the balance of the skins microflora is upset, adding to inflammation and disease flares[1]. Skin surface pH has been found to be higher (less acidic) in children with eczema, in both affected and unaffected areas of skin. This is likely caused by problems with the elements needed to form the acid mantle, and so is part of the process leading to the development of eczema rather than a result of the condition[4].

Cold Weather and Eczema

Climate and weather conditions affect how common eczema is in any part of the world. Scientific research studies observing thousands of children from all over the world have indicated that eczema is more common at northern latitudes, at areas with lower temperatures, more rainfall, more use of central heating and less sunshine[5,6,7]. It seems to be less common in areas that are warmer, areas with more sunshine and low use of central heating[6,7]. In particular, the combination of lots of sunshine and high temperature seems to have a protective effect against eczema, while combined high humidity, high rainfall and low sunshine seems to be associated with more cases of eczema[6].

> ❖
>
> Eczema is more common at northern latitudes, with lower temperatures, more rainfall, more central heating use and less sunshine[6,7]
>
> ❖

Climate and weather may trigger eczema in at-risk children, with cases of previously unaffected children developing eczema when they move to northern latitudes from warmer parts of the world[6].

The impact that weather has on the skin can change depending on the combination of weather elements such as humidity, temperature and sunlight. The extremes of exposure tend to be harmful, for example both too much and too little humidity can cause problems for the skin barrier. While low temperatures harm the skin barrier, high temperatures lead to more sweating, which can have an irritant effect on the skin. Humid conditions worsen the effect of high temperature, as sweat then struggles to evaporate. Probably related to this, a study of 5595 children from the United States found that high temperatures, high daily sun exposure and high humidity were linked to poorly controlled eczema[5].

A sweet spot in climate variables like UV, humidity and temperature probably exists and there will also be variations between individuals in terms of the conditions they feel most comfortable in. During cold weather at northern latitudes, lowering the indoor heating temperature by a few degrees, adjusting indoor humidity levels and spending more time outdoors where the air is less dry may help to mitigate some weather-associated triggers[6].

Allergy and the Skin Barrier

Having a leaky skin barrier increases the risk of allergens getting through the skin to meet the immune system, including dust, pollen and food proteins. A Canadian study found that 1 year old children with both eczema and contact allergy were at greater risk of developing food allergy by the age of three than those who had just one or the other[8].
Food allergy is associated with longstanding and more severe eczema. Research has indicated that eczema frequently appears before the development of food allergy, suggesting that it may be involved in causing allergy to develop[9]. It is thought that these food allergens can enter through the skin leading to the immune system later recognising them across the

gut when they are eaten. This theory is supported by the observation that food allergies can develop without the child having ever eaten the culprit foods, particularly egg and peanut[9].

There are various factors and genetics at play in the development of food allergy and the skin barrier is not the only pathway through which an allergy may develop, but research supports the idea that regularly treating the skin barrier with moisturiser, appropriate cleansers and prescribed treatments to maintain its integrity may alter the course of allergic disease, as well as improving the state of the skin.

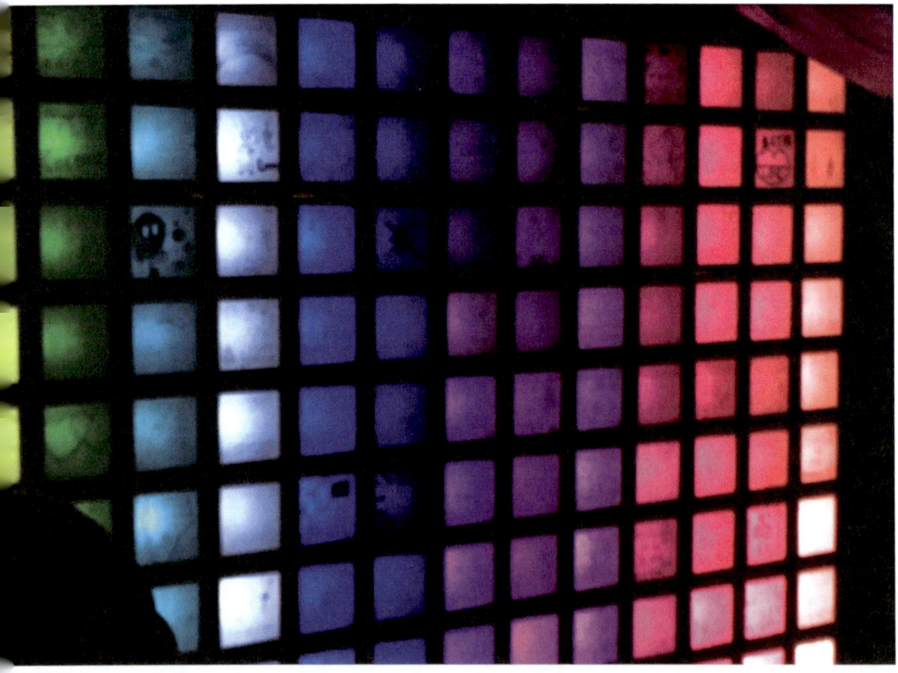

Maintaining the Barrier

Regular use of moisturiser is very important in maintaining the integrity of the skin barrier. Regular use of therapeutic moisturisers from infancy has been linked to better skin barrier function, improvements in itch and reduction in the frequency of flares and less overgrowth of unwanted skin bacteria[1,3]. Regular moisturiser has also been linked to a reduction in the chance of eczema appearing in newborns at high risk of eczema[1,3]. Prescribed medical treatment should also be followed to resolve flares and reduce how often they occur.

Getting effective skincare onto the children that need it is important as they don't know to do it themselves. It can make such a difference to their symptoms, the visible state of their skin and their general wellbeing. They are unlikely to be entirely happy about it and finding ways around this is part of the parenting role, but no one is saying it's easy.

❖

Regular use of moisturiser in newborns at risk of eczema can reduce the chance of it appearing[1,3]

❖

Parents do a great job in often difficult negotiating situations. It is important for skincare not to develop into a stressful time for the child or family, as it is likely to be a part of life for a long time. Children's opinions change and it can be frustrating when a reliable coaxing technique suddenly stops working. But try not to get disheartened, new methods can usually be found and children do also get older and wiser.

How often to bathe children with sensitive skin and what to put in the bath is often debated with varying opinions and recommendations on how often to do it and what to use[3,9]. Keeping a child clean is important for the health of their skin, for their general health, their social development and to prevent spread of infectious disease. Cleansers can vary significantly in pH and alkaline preparations such as soap can damage the skin barrier by increasing the pH at the skin surface[4].

> ❖
> Keeping a child clean is important for the health of their skin, for their general health, their social development and public health
> ❖

As discussed in Chapter 2, many skin oils and dirt molecules tend not to dissolve in pure water, as they are hydrophilic (water-hating). Cleansers are designed to remove these particles by containing surfactants, which are molecules that attract these water-hating substances and allow them to be rinsed off. Unfortunately however, surfactants cannot tell the difference between unwanted oils and the oily lipids and proteins in the skin barrier and can end up removing varying amounts of those too, leading to weakening of the skin barrier, dryness and irritation.

There are many types of surfactant and they all strike a balance between skin cleansing and skin irritation. Anionic surfactants such as true soap and some synthetic detergents (syndets) like sodium laureth sulphate are the most irritant. Surfactants also affect the pH of a formulation, with those containing soap typically having an alkaline pH, whereas syndets are generally neutral or acidic[4].

Gentle, moisturizing, syndet cleansers should be used to limit damage to the skin barrier while cleaning the skin[3]. Also, avoiding heavily fragranced products in a child with irritated skin reduces the chance of these molecules crossing the skin barrier and causing a contact allergy to fragrance.

Using appropriate cleansers, infants and children can be bathed daily if needed, with one school of thought suggesting limiting a bath to 10 minutes or less, using comfortable but lukewarm to warm water, as hot water encourages itchiness and scratching.

Gentle, moisturising syndet cleansers should be used to limit damage to the skin barrier while cleaning the skin[3]

Snot

The inside of our mouth, nose, sinuses, throat, lungs and gut is lined with mucus, produced by special mucus producing cells. When it comes out of our nose we call it snot.

Mucus is a permanent protective coating over these tissue surfaces, preventing them from drying out, and it also traps unwanted guests like bugs, dust and other particles before they can make it further into our bodies. It contains antibodies that allow the immune system to recognize invading bugs and viruses, enzymes to destroy them and proteins and other cells to keep it sticky, gooey and snotty[10]. Mucus production increases in response to large amounts of microbial invaders, such as during a cold or viral infection, or if allergens are present.

Mucus is a protective coating, preventing tissues from drying out and trapping unwanted guests

Snot lying on skin can cause irritation through both the constant dampness at the area and the frequent nose wiping or blowing, which can further irritate and break the skin in the area. Baby wipes are really convenient when out and about. They are soft compared to some forms of tissue and manage to wipe away a whole range of muck. A study on newborn babies in the first four weeks of life showed that wipes did not produce any more skin irritation or problems with dryness than cotton wool and water, but this trial was supported by the company making the wipes[2].

However, when wiping skin that is exposed to cold weather, baby wipes leave the skin damp with no protective cover. Many wipes contain mild surfactants that may cause irritation if left on the skin. For drippy noses, try using a soft dry tissue to gently wipe, followed by a quick smear of bland ointment across the nose to give a protective, barrier-repairing coating. This is a better option for keeping the skin healthy or encouraging healing of irritated skin.

❖
A soft but dry tissue followed by a quick smear of bland ointment gives a protective, barrier-repairing coating
❖

Saliva

Saliva is incredibly important for the health of our mouth and our health in general. The mucus in saliva helps bind chewed up food into an easily swallowed slippery ball; salivary enzymes break up the molecules in food so that we can actually taste them and kill off unwanted bacteria; and saliva essentially flushes our mouth, washing away food and muck from our teeth and tissue crevices that would otherwise stay and rot[11]. Yum.

While most of the time adults manage to keep saliva inside their mouths, kids manage to get it everywhere. Lip lickers splash it about, dummy-users smoosh it up onto their cheeks, finger suckers slobber it over their hands and droolers let it dribble onto their necks, chests and beyond. A teething child with a cold is a truly miserable sight.

When the supply of childish saliva is going strong, the skin is pretty much permanently wet, and irritation can appear because of this constant wetness and the enzymes the saliva contains. Stopping a drooler from drooling just isn't practical or possible, so back-up protection techniques have to be found.

Dribble bibs are invaluable for younger children, as long as they're changed once the area around the neck is wet. As far as possible try not to wipe up the dribbles and puddles with wet wipes, as these tend to leave the skin wet and unprotected and may leave mild surfactants on the skin like earlier. Try to use soft tissues, then as soon as the skin is dry apply a quick smothering of a bland ointment moisturiser as a protective and barrier repairing coating, over the lips / cheeks / nose. The same can also be used on chest or neck.

Fingers are tricky. Finger suckers suck their fingers for good reason and do not intend on giving it up easily. Constantly wet fingers can start to break down through maceration of the tissue (like being in the bath too long) and the irritant properties of saliva, increasing the chance of infection. Many kids' hands dry out and crack in cold weather, in the same way as adults do, and adding drooly wetness onto that can lead to problems.

> ❖
> Stopping a drooler from drooling isn't practical or possible
> ❖

Also, bland ointment based moisturisers aren't designed to be eaten. They are used in lip balms, and small amounts are very likely ingested in that way and do us no harm, but repeatedly coating little fingers in petroleum jelly for them to be put back into mouths would not be advised. Edible substances have been used in skincare but are quite likely to encourage finger sucking!

You may find that a finger sucker with sore fingers picks alternative fingers to suck, and if that happens you can get to work drying and moisturizing the rejected digits. Another option is approaching the deeply sleeping child, applying a coating of moisturizer to the fingers then, and adding some mitts on top if possible. That at least allows the barrier to repair somewhat overnight.

Coaxing the Stubborn

Getting creams onto children can be tricky. Here are just some ideas for ways to make it a happier experience.

Warm it up

No one likes cold cream. Despite full knowledge and decision-making capacity many adults won't apply cold cream to their skin, so why should your toddler put up with it. In skin clinics it often feels like we ask parents to attempt the impossible, applying cream all over a small child several times a day. Trying to do this every day for years can become depressing and stressful.

Being cold is one of the worst things that can happen to a small child, so they would have you believe anyway. So try not to undress them all at once, do an area at a time, in a warm room, warming the cream in your hands before slathering it onto them. Put their clothes or pyjamas on the radiator so there's something nice and warm to go on top.

Travel Kit

Relatively simple - put a small supply of cream and ointment moisturisers and a packet of soft tissues in every bag that travels with you and your child, then it's there when you need it.

Messy Play with Moisturiser

Many kids like to get elbow deep in sticky stuff. Why not try them with something that might help their skin while they're at it. Sure, moisturiser only comes in white or clear forms, but squirt some onto coloured plastic or card and see if they play with it. Cover some plastic toys with it and show them how slippery they get. Rub it up their arms while they're at it, and try and get their feet into it too.

While You Were Asleep

30-40 minutes after falling asleep, many kids out of infancy (even the lightest sleepers) are in a bit of a coma. Various things can be done at this time that would be whole-heartedly rejected when awake. Nail cutting can be safer as a nocturnal activity than trying to wrestle a wriggling foot near to sharp scissors. Gently and tactically applying warmed moisturiser to any available skin can also be attempted. Use cold cream at your peril.

Favourite Characters

Advertisers have played on the power of persuasion in small children for decades. You can use this too. Sure, the cartoon character moisturizing cream might contain fragrance and few active ingredients, and the therapeutic stuff comes in a plain bottle in comparison, but what is to stop you from cutting characters out of a magazine and sticky-back plasticking them onto the plain bottle. Super Flying Hero Cream is much more exciting than a standard bottle.

Along the lines of favourite characters, children love to copy, whether it's eating daddy's breakfast, wearing grandpa's shoes or having the same cup as their big cousin. Letting your child see you applying moisturizing cream to yourself is likely to be more encouraging for them than having to go it alone.

Summary

- The effects of cold weather, snot and saliva cause irritation, damage and disruption to the skin barrier. Clean and dry skin with soft, dry tissues and protect it with a smear of bland ointment.
- Eczema, a chronic skin condition related to problems with the skin barrier, is common in children and more common at northern latitudes, where there is more rainfall, colder temperatures and more use of central heating.
- Regular use of moisturiser in newborns at high risk of eczema can reduce the chance of it developing.
- Hacks like warming creams, disguising bottles and incorporating creams in play may help to prevent skincare from becoming a stressful time for the child and family.

References

1. Kim BE, Leung DYM. Significance of skin barrier dysfunction in atopic dermatitis. Allergy Asthma Immunol Res. 2018;10(3):207-215
2. Lavender T, Furber C, Campbell M, Victor S, Roberts I, Bedwell C, Cork MJ. Effect on skin hydration of using baby wipes to clean the napkin area of newborn babies: assessor-blinded randomized controlled equivalence trial. BMC Pediatr. 2012;12:59
3. Telofski LS, Morello AP, Mack Correa MC, Stamatas GN. The infant skin barrier: can we preserve, protect and enhance the barrier? Dermatol Res Pract. 2012;2012:198789
4. Ali SM, Yosipovitch G. Skin pH: from basic science to basic skin care. Acta Derm Venereol. 2013;93(3):261-267
5. Sargen MR, Hoffstad O, Margolis DJ. Warm, humid and high sun exposure climates are associated with poorly controlled eczema: PEER (paediatric eczema elective registry) cohort 2004-2012. J Invest Dermatol. 2014;134(1):51-57
6. Silverberg JI, Hanifin J, Simpson EL. Climatic factors are associated with childhood eczema prevalence in the United States. J Invest Dermatol. 2013;133:1752-1759
7. Weiland SK, Husing A, Strachan DP, Rzehak P, Pearce N. Climate and the prevalence of symptoms of asthma, allergic rhinitis, and atopic eczema in children. Occup Environ Med. 2004; 61(7):609-615
8. Tran MM, Lefebvre DL, Dharma C, Dai D, Lou WYW, Subbarao P, Becker AB, Mandhane PJ, Turvey SE, Sears MR, Canadian Healthy Infant Development Study Investigators. Predicting the atopic march: results from the Canadian Healthy Infant Longitudinal Development Study. J Allergy Clin Immunol. 2018;141(2):601-607
9. Tsakok T, Marrs T, Mohsin M, Baron S, du Toit G, Till S, Flohr C. Does atopic dermatitis cause food allergy? A systematic review. J Allergy Clin Immunol. 2016;137(4):1071-1078
10. Watson S. The truth about mucus [Internet]. United States of America: WebMD.com; [updated 2014; cited 2019 June]. Available from: https://www.webmd.com/allergies/features/the-truth-about-mucus#1
11. Tiwari M. Science behind human saliva. J Nat Sci Biol Med. 2011;2(1):53-58

8. All in a Flurry: Making Time

8. All in a Flurry:
Making Time

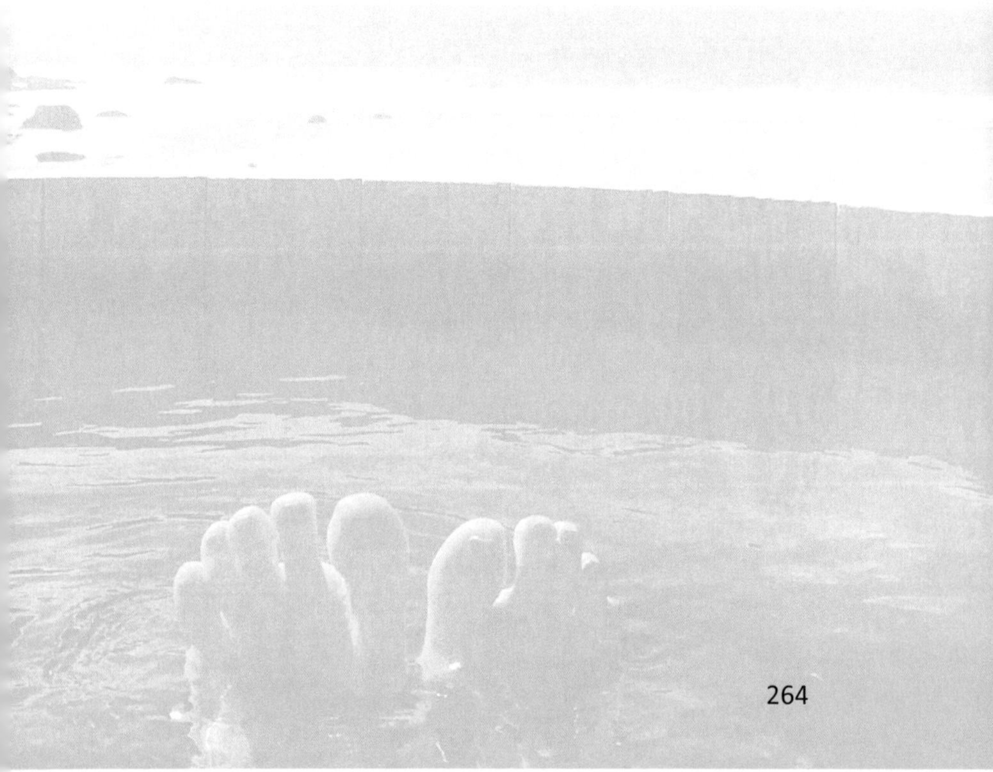

Time flies. We fly through seasons, the colours in our lives going from blues and yellows to browns and golds to silvers and sparkles then greens and whites and all the way back to the start again, repeated forever. We hang onto precious habits and routines to anchor ourselves somewhere in this spinning world, but sometimes years later we find that these securities have not cared for us as much as we have for them.

Many skincare habits that we nurture will start to show in our skin, especially in our faces and hands. If we can only bag some time in our busy lives to pin down healthy habits early this can make a great difference when suddenly we find it is 20 years later and the fashions we grew up with are back in the shops. Good skincare does not need to cost a fortune or be complicated, just well chosen and regularly used.

Good skincare does not need to cost a fortune, or be complicated

Organising the Cupboard

It is so easy to get distracted by the variety of skincare on shop shelves and end up with two or three face washes, four hand creams and several expensive serums. Organising our products by their function and just replacing what is needed can help to avoid indiscriminate buying based on sales recommendations or packaging. Give products a chance and avoid jumping from one to another on a daily basis, as this will get the best out of them for your skin and also help you recognise what works best for you.

Look at who makes your skincare. Is the company producing your regular moisturiser wholly committed to skincare? Or are they a big brand with a sideline in it? Skincare is a multimillion-pound industry with everyone from make-up brands, to home ware companies, to celebrities wanting a section of the market. The companies dedicated to skincare tend to research and develop their products and are most likely to produce the best quality goods. Good choices can generally be found in the skincare (not make-up) aisles of larger pharmacies and specialised retailers.

So many of us buy skincare products based on recommendations from magazines and blogs, or because 'it worked for such-and-such and she/he has great skin'. As much as we want it to be, skincare just isn't a one-size-fits-all issue. What works for others may not work for you, and - as is the point of this book - what works best for your skin is likely to vary from time to time depending on changeable factors including season and weather. However, in terms of basic skincare and what to start with in the cupboard, a suggestion for a toolkit would be:

Daily Care:
Gentle syndet facial cleanser, syndet body wash
Daytime facial lotion (cream for dry skin) +/- serum
Facial night cream
Hand and body cream
Foot cream
Overnight ointment for hands and other dry areas

Once or twice a week:
Nighttime facial, ingredients depending on skin type
Overnight retinoid, can be used more often if tolerated
Facial mask, ingredients depending on skin type
Facial exfoliating brush (use with care, or avoid, in rosacea and irritated skin)
Sand or salt body scrub
Foot file or pumice stone

Once these are unpackaged and to hand in the bathroom or shower, using them takes minutes. Facemasks don't need to be a morning's work, if time is limited some preparations are thick enough to be applied at the start of a shower and washed off at the end.

As with anything, the biggest struggle is often making a change to a situation; choosing to buy a gentle, moisturising wash over a pure soap, changing where you buy your skincare from, choosing to add a few extra bits into your daily routine. But if you are prone to skin problems than taking these steps is worth it.

Dry face

The face is particularly exposed to the effects of the weather and can be one of the first areas to noticeably dry out in the colder months. Dryness at the face feels rough to touch, sandpapery and can be itchy.

Make sure that your facial moisturiser is appropriate for your skin type. Many products designed for oily skin can be very drying if used daily on normal or combination skin. Many people need to use a different facial moisturiser in summer and winter, consider this if you are moisturising everyday with a lotion that worked well in July but isn't quite cutting it in October. Try something thicker, containing lipids and barrier repairing ingredients.

Cleansing. The commonly used phrase 'it left my skin squeaky clean' usually means 'my skin feels tight and dry'... It implies that the product has had a good eat away at the lovely lipids and molecules making up your skin barrier. Wash your face with a gentle syndet or soap-free cleanser morning and evening. Consider moisturising cleansers like cleansing milks, lotions and balms, but beware of the potential for comedogenicity with some of these.

Take care with foam or scrub cleansers, these have more potential to strip away your skin barrier lipids and cause physical irritation. Micellar waters and other leave-on cleansers can be great for cleaning dry faces, but like wet wipes they contain mild surfactants and preservatives that can irritate the skin if left on, so consider rinsing your skin after using them. Apply a moisturiser afterwards.

Night creams and hydrating masks can give a welcome boost. While technically an antioxidant, nighttime retinol or retinoid creams can also help reduce dryness and wrinkling in the skin. They can irritate, so try overnight once a week to start with and work up from there if you can.

Dry hands

Keep a portable tube of fragrance free, therapeutic moisturising cream with you at all times. Put some in every room with a sink and keep some in your bag. Reapply when needed and after washing your hands. Replace all hand soaps with gentle, moisturising syndet hand washes. Dry your hands properly after washing, especially around the nails. Get a tub of thick ointment moisturiser and multiple pairs of cotton gloves and apply the moisturiser under cotton gloves at night a few times a week, or every night for a while if needed. If your hands are persistently red and itchy then a prescription for steroid cream may be needed to calm them down.

Dry feet

Feet are often stuffed into socks and boots and neglected during cold weather. They can dry out dramatically, leading to painful cracks especially at heels and around the toes. Take special care of your feet. Get a good thick foot cream and plaster them everyday. Use a foot file or pumice to rub the hard bits away. Make sure your boots fit and aren't rubbing away at your feet. If you fancy a bit extra, investing in an inexpensive home foot spa can be a welcome comfort after a day in the cold. Dry your feet completely and then apply moisturiser afterwards. Bliss.

Red nose

Got a runny nose? Regular trauma from tissues will lead to redness and dryness at the nose. Keep going with the tissues but after blowing apply a thin coat of ointment moisturiser here as often as possible. Cover the area in ointment when you're in the house and not bothered about shine.

Not got a cold? Little can be done if this is a physiological response to cold like cold induced vasodilatation, but if other triggers set it off, or if it is persistent and occasionally worsens then try treating it as rosacea. Avoid any triggers as far as you can. Try the skincare advice in Chapter 3, not forgetting to use a facial SPF every day. See your doctor for their opinion and to try any medical treatment if necessary.

Dry lips

Our lips are exposed to trauma from the weather, saliva and the foods we eat. Lips have a thin stratum corneum with lower levels of ceramides and less mature keratinocytes than other body areas, leading to a weakened skin barrier with higher rates of water loss across the epidermis. There are also no sebaceous glands in the lips, meaning no protective sebum coating on top of the skin. Because of all this lips can easily dry out, becoming rough, cracked, chapped and painful[1,4].

Lip balms usually contain occlusive agents, designed to form a protective film on the skin surface. Hunt down a lip balm or lip treatment that also contains ceramides, or compounds known to help in the formation of ceramides like lactic acid, which can support the skin barrier. Using a lip balm with an SPF will protect against the harmful and ageing effects of UV light.

For Skin that Sparkles

Our arms, legs, back and front also suffer the effects of being covered up in cold weather. Indulging in a bit of extra skincare can boost the effect of getting dressed up for any event during cold weather. There are heaps of places offering body scrubs, wraps and exfoliating treatments, but if there's not the time or budget for it try the quick fix below. Also, sometimes you can produce a better result with a few treatments at home than one expensive one at a salon.

All you often need is a salt or sand body scrub (available in supermarket) and a good therapeutic moisturising cream with a combination of active ingredients and some additional ingredients (larger pharmacy or specialised retailer). Fake tan or shimmery cosmetics can be used on top of these as desired, but avoid using any products that you haven't tried before within 48 hours of an event, just in case they don't agree with you.

Day -7:	Start moisturising all over, AM & PM. Use a moisturising syndet bodywash for washing.
Day -6:	Moisturising AM & PM.
Day -5:	Body scrub in AM, moisturise AM & PM.
Day -4:	Keep moisturising.
Day -3:	Keep moisturising!
Day -2:	Repeat scrub, continue moisturising
Day -1:	Keep.Moisturising
Day 0:	Wash, moisturise all over, enjoy!

Using a body scrub too often may irritate the skin, but it does encourage desquamation of the dull, rough keratinocytes at the skin surface, giving a smoother surface that can be enhanced further by adding a good moisturiser on top. If you use fake tan, try applying it on the evening of day -5, then topping it up on the evening of day -2 or day -1. I have not yet managed to find a fake tan that doesn't have some smell to it so I try to avoid applying it on the day or the night before an event.

The above plan might look very simple, but this is the beauty of skincare – it really doesn't have to be complicated to produce great results. Science and research are constantly advancing and we will all benefit from groundbreaking discoveries as they occur, but in the meantime you are not missing out at all by consistently using straightforward, well-chosen products.

Summary

- Skincare does not need to be complicated or expensive to be effective.
- Buy your skincare from companies that are dedicated to skincare. Be organized and try to avoid impulse buys.
- Start reading the ingredients list of skincare products and buy based on these, not packaging or adverts.
- Be prepared to adapt your skincare regime with the seasons, depending on the look and feel of your skin.

References

1. Tamura E, Ishikawa J, Naoe A, Yamamoto T. The roughness of lip skin is related to the ceramide profile in the stratum corneum. Int J Cosmet Sci. 2016;38(6):615-621
2. Isoda K, Nakamura T, Yoshida K, Tamura E, Atsuta N, Ishida K, Takagi Y, Mizutani H. The efficacy of a lip balm containing pseudo-ceramide on the dry lips of sensitive skin-concious subjects. J Cosmet Dermatol. 2018;17(1):84-89

Acknowledgements

Thank you to my parents for the opportunities they gave me and encouraged me to take, and to my husband and our children for their support and generally being brilliant. Thank you to Catherine Davidson Photography for the cover photo, and to Cambridge Scholars Publishing for giving me the opportunity to write this book.